Disrupted Cities

Whether we like it or not, life on a globally integrated and rapidly urbanizing planet involves intensifying dependence on vast systems of infrastructure. These sustain the mobility, health, energy, water, food, and waste removal services without which modern city life is all but impossible. When such infrastructures work they are often ignored or taken for granted, at least by those lucky enough to have good and continuous access. Paradoxically, however, when they collapse or are disabled they are at their most visible. During such disruptions, urbanites momentarily become experts in the amazingly complex and multiscaled infrastructures that they rely on to sustain their every moment.

Disrupted Cities is the first book to explore the ways in which reactions to, and experiences of, the collapse of infrastructures within and between cities are constructed, imagined, represented, and contested. Focusing primarily on case studies of iconic moments of infrastructural disruption within North America, the book shows that such disruptions unerringly work to expose the hidden politics of contemporary urban life.

Stephen Graham is Professor of Human Geography at Durham University in the UK. He has a background in urbanism, planning, and the sociology of technology. His research addresses the complex intersections between urban places, mobilities, technology, war, surveillance, and geopolitics. He is Academic Director of the International Boundaries Research Unit (IBRU) and Associate Director of the Centre for the Study of Cities and Regions (CSCR), both at Durham. His books include *Telecommunications and the City*, *Splintering Urbanism* (both with Simon Marvin), *The Cybercities Reader*, and *Cities, War and Terrorism*. His latest book, *Cities Under Siege: The New Military Urbanism*, will be published by Verso in 2009.

D1418819

Disrupted Cities
When Infrastructure Fails

Edited by

Stephen Graham
Durham University, UK

Routledge
Taylor & Francis Group

NEW YORK AND LONDON

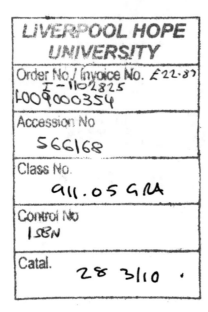
First published 2010
by Routledge
270 Madison Ave, New York, NY 10016

Simultaneously published in the UK
by Routledge
2 Park Square, Milton Park, Abingdon, Oxon OX14 4RN

Routledge is an imprint of the Taylor & Francis Group, an informa business

© 2010 Taylor & Francis

Typeset in Sabon and Trade Gothic by EvS Communication Networx, Inc.
Printed and bound in the United States of America on acid-free paper by Edwards Brothers, Inc.

Library of Congress Cataloging in Publication Data
Disrupted cities : when infrastructure fails / edited by Stephen Graham.
p. cm.
1. Emergency management. 2. Disasters—Social aspects. 3. Infrastructure (Economics.) I. Graham, Stephen.
HV551.2.D59 2009
363.34—dc22
2009009663

ISBN10: 0-415-99178-1 (hbk)
ISBN10: 0-415-99179-X (pbk)
ISBN 10: 0-203-89448-0 (ebk)

ISBN13: 978-0-415-99178-0 (hbk)
ISBN 13: 978-0-415-99179-7 (pbk)
ISBN 13: 978-0-203-89448-4 (ebk)

For Frank

CONTENTS

ACKNOWLEDGMENTS

I would like to thank Simon Marvin and the staff at SURF, Salford University, for key inputs into the early stages of this book. Thanks also to the authors for working so hard to produce chapters which address the book's theme so specifically; to Nigel Thrift, for collaboration on earlier work which helped greatly in formulating the book's introduction; and to Stephen Rutter and Leah Babb-Rosenfeld at Routledge, New York, for their support and patience in the production of this book.

Deb Cowen would like to thank Steve Graham, Simon Marvin, and Neil Smith for helpful comments on an earlier draft of chapter 5. She would also like to thank the Centre de Cultura Contemporania de Barcelona, Louise Amoore, and Steve Graham for convening "Targeted Publics: Arts and Technologies of the Security City," as well as the participants of this workshop for such productive dialogue on these themes. She also thanks Byron Moldofsky in the Cartography Lab at the University of Toronto for adapting the DHS map on very short notice. A postdoctoral fellowship from the Social Science and Humanities Research Council of Canada made this research possible.

The longer version of Tim Luke's chapter was originally prepared for the Conference on Urban Vulnerability and Network Failures: Constructions and Experiences of Emergencies, Crises, and Collapses, April 28–30, 2004. Various excerpts from it appear in: "Technology as Metaphor: Tropes of Construction, Destruction, and Instruction in Globalization." *Metaphors of Globalization: Mirrors, Magicians and Meanings*, ed. Vincent Pulpit, Nita Shan, Markus Kornprobst, and Ruben Zaiotti. New York: Palgrave Macmillan, 2008; "Unbundling the State: Iraq, the 'Recontainerization' of Rule, Production, and Identity." *Environment & Planning A*, 39 (2007); and, "Finding New Mainstreams: Perestroika, Phronesis, and Political Science in the United States." In *Making Political Science Matter: Debating Knowledge, Research, and Method,* ed. Sanford Schram. New York: New York University Press, 2006.

Stephen Graham's chapter is an updated and altered version of a paper published in the journal *City*, Volume 9, Part 2. Various excerpts of the chapter also appear in *Cities Under Siege: The New Military Urbanism*. London: Verso, 2009; "Demodernizing by Design: Everyday Infrastructure and Political Violence." In *Violent Geographies: Spaces of Terror and Political Violence,* ed. Alan Pred and Derek Gregory. New York: Routledge 2006; and "Urban Metabolism as Target: Contemporary War as Forced Demodernisation" in *In the Nature of Cities: Urban Political Ecology and the Politics of Urban Metabolism,* ed. Nik Heynen, Maria Kaika, and Eric Swyngedouw. London: Routledge, 2005.

Simon Marvin and Will Medd's chapter is a shortened, revised, and up-dated version of Simon Marvin and Will Medd, "Metabolisms of Obecity: Flows of Fat through Bodies, Cities, and Sewers." *Environment and Planning* A 38, no. 2 (2006): 313–324.

Stephen Graham

PREFACE

It is one of the great paradoxes of urban life in the Global North that it often requires the collapse of the great, stretched-out infrastructures that sustain the city—the power grids, transport networks, water systems, and digital circuits—for their critical importance to become manifest. Infrastructures—the largest urban architectures of all—often become most visible when they lie dormant or inactive—temporary ruins to the dreams of modernity, mobility, and circulation that underpin them.

Infrastructure disruptions bring fleeting visibility to the complex practices and technologies, stretched across geographic space, that continually bring the processes of urban life into being. When the lights are out, the server down, the airport closed, or sewer system blocked, the material networks which continually work to produce urban life that is usually rendered "normal" quickly become a public obsession. Obscure technical artifacts and processes fill the newspaper pages; distant networks are momentarily mapped and rendered visible in the media. The politics, and dramatic inequities, of the vast technological circuits of our world are momentarily forced to the foreground.

Infrastructure disruptions radically transform urban life as urbanites seek to adjust, and cope with, uncanny worlds of darkness, cold, immobility, hunger, thirst, or dirt. Often, the sense of crisis and the search for alternatives helps forge new ideas about what urban life might be.

In many Global South cities, conversely, infrastructure disruptions tend always to be in the foreground. In such burgeoning cities, the politics of urban life are played out through unequal struggles to improvise reliable water, power, transport, or communications when such services have rarely reached levels of access or reliability that allow them to be taken for granted or rendered largely invisible.

Disrupted Cities is the first book to peer into the complex processes and politics that surround infrastructure disruptions and failures. Using the latest advances in social and urban theory, the book draws out the implications of collapses in the taken for granted circulations of cities for our understanding of contemporary urban life. Drawing together leading researchers from geography, political science, sociology, and public policy, the book explores some of the iconic infrastructure disruptions that have been such a feature of urban life in the past decade— the 2003 blackout in the U.S. Northeast, the 2005 Hurricane Katrina disaster in New Orleans, and the 2001 SARS crisis. These are combined with discussions of less visible infrastructure disruptions, such as the clogging of city sewers by discarded fats or the continual breakdown of services in many Global South cities. The book also explores how state and nonstate political violence now amplify themselves by deliberately attacking the infrastructural circuits of cities and how discourses of urban security are manipulating such threats to radically remodel and securitize urban life.

A cutting-edge, deeply topical, and highly interdisciplinary book, *Disrupted Cities* will be invaluable for advanced undergraduate and postgraduate teaching across the full range of urban disciplines: in urban geography and sociology; urban politics, planning, and architecture; and urban environmental studies and public policy. In centering on infrastructure and its disruption, which necessarily lie at the interface of technology, politics, nature, and culture, the chapters in this book will be highly stimulating for urban scholars, students, and researchers, whatever their disciplinary starting points.

CHAPTER 1

When Infrastructures Fail

Stephen Graham

The town exists only as a function of circulation and of circuits; it is a singular point on the circuits which create it and which it creates. It is defined by entries and exits: something must enter it and exit from it.[1]

DISRUPTED CITIES

On our rapidly urbanizing planet, the everyday life of the world's swelling population of urbanites is increasingly sustained by vast and unknowably complex systems of infrastructure and technology stretched across geographic space. Immobilized in space, they continually bring into being the mobilities and circulations of the city and the world. Energy networks connect the heating, cooling, and energizing of urban life through infrastructure to both far-off energy reserves and global circuits of pollution and global warming. Huge water systems sate the city's insatiable thirst, their waste water and sewerage parallels removing human and organic wastes from the urban scene (at least partially). Within cities, dense water, sewerage, food and waste distribution systems continually link human bodies and their metabolisms to the broader metabolic processes through which attempts are made to maintain public health. Global agricultural, shipping, and trade complexes furnish the city's millions with food. Highway, airline, train, and road complexes support the complex and multiscaled flows of commuters, migrants, tourists and refugees, as well as materials and commodities, within and through the global urban system and its links with hinterlands and peripheries. And electronic communications systems provide a universe of digitally mediated information, transaction, interaction, and entertainment which is the very lifeblood of digital capitalism and which is increasingly assembled based on assumptions of always being "on." The vital material bases for cyberspace are largely invisible and subterranean. They also link intimately both to the electrical infrastructures which allow it to function, and to the other infrastructural circuits of the city as they themselves become organized through digital media.

Whilst sometimes taken for granted—at least when they work or amongst wealthier or more privileged users and spaces—energy, water, sewerage, transport, trade, finance, and communication infrastructures allow modern urban life to exist. Their pipes, ducts, servers, wires, conduits, electronic transmissions, and tunnels sustain the flows, connections, and metabolisms that are intrinsic to contemporary cities. Through their endless technological agency, these systems help transform the natural into the cultural, the social, and the urban.

Infrastructural edifices thus provide the fundamental background to modern urban everyday life—a background that is often hidden, assumed, even naturalized. They fundamentally underpin the ceaseless and mobile *process* of city life in a myriad of ways. This process inevitably works across many geographical scales, from the level of the human body and its metabolisms—through which the food, water, and energy brought to the city through infrastructural circulation actually flow—through the city, region, and nation to the transnational and even planetary—with its transnational networks of energy extraction and flow, airline travel, electronic communication, food trade, port systems, and the movement of solid, liquid, and gaseous wastes. Much-debated processes of globalization—beneath the fast-fading hyperbole of the business press—rely after all on vast and unimaginably complex material circuits of infrastructure within which cities invariably act as the dominant hubs of built networks, the predominant centers of demand (for energy, food, water, transport, and communications), and the dominant centers for generating pollution and waste of all forms.

The political, economic, social, and environmental importance of the world's lattices of urban infrastructure can only grow as the world becomes more urban. Well over 50 percent of the world's population lives in cities; 75 percent of the world's population of over 9 billion people is projected to live in them by 2050.[2] Within just over four decades there will be fully 7 billion people living in the world's cities, 4.75 billion more than in 2007. The overwhelming majority of these will be in the burgeoning cities and megacities of the so-called developing world: in Asia, Africa, and Latin America.

As this great demographic and geographic shift continues, humankind will become ever more reliant on functioning systems of urban infrastructure. Indeed, the very nature of urbanization means that every aspect of people's lives tends to become more dependent on the infrastructural circuits of the city to sustain individual and collective health, security, economic opportunity, social well-being, and biological life. Moreover, because they rely on the continuous agency of infrastructure to eat, wash, heat, cook, light, work, travel, communicate, and remove dangerous or poisonous wastes from their living place, urbanites often have few or no real alternatives when the complex infrastructures that sometimes manage to achieve this are removed or disrupted.

This book is the first to concern itself exclusively with what happens when the infrastructural flows or metabolisms of the modern city, which so often come to be considered so normal that urbanites may even come to see them as culturally banal, invisible, even boring, are suddenly interrupted or disturbed. In what follows we are concerned with what happens when technical malfunctions, interruptions in supplies of resource, wars, terrorist attacks, public health crises, labor strikes, sabotage, network theft, extreme weather, and other events usually considered to be "natural" (floods, earthquakes, tsunami, etc.) disrupt the flows of energy, water, transportation, communication, and waste that are the very lifeblood of the contemporary city. To address this question, *Disrupted Cities* brings together an unprecedented collection of cutting-edge research from an influential range of academic writers within geography, political science, sociology, and urban planning. These explore a series of instantly iconic case studies of such disruptions with a high degree of sophistication and detail. Such cases are drawn largely, but not exclusively, from North America and include very high profile events such as the September 2001 attacks on New York and Washington, the 2003 power blackout in the North Eastern seaboard, the devastation of New Orleans by Hurricane Katrina in 2005, and the infiltration of the global airline system by the SARS virus in 2003.

Disrupted Cities, however, is not just an exploration of iconic recent infrastructural disruptions that have been such a feature of global urban life in the early twenty-first century, notably in the United States (although we must be conscious that global media biases mean that infrastructure disruptions beyond North America, Europe, Australasia, and East Asia rarely receive much in the way of news coverage within mainstream Western media). The book complements these studies with analyses of much less known but equally important infrastructural disruptions: the deliberate targeting of city infrastructures in Iraq and the Occupied Territories by the U.S. and Israeli state militaries; the hidden sclerosis of city sewers caused by discarded fats; the ways in which concern about disruptions is being used to politically reorganize and securitize global port systems in the wake of the "global war on terror"; and the normal disruptions of city infrastructure that tend to characterize life in the burgeoning megacities of the Global South—at least for the populations of informal settlements.

Together, these diverse, critical, and connected analyses provide the first systematic social scientific study of practicalities, politics, and implications of events where the circuits sustaining flows of energy, water, waste, sewerage, transportation, or communications within or between cities break down, are deliberately attacked, or become infused with malign infiltrations. We reveal in unprecedented detail how the processes of the politics of city life respond to such crises, how the threats of disruption can be manipulated for political ends, and what such events tell us about the times and geographies when infrastructural flows and metabolisms sustaining urban life continue as "normal." Above all, the book raises major questions about the challenges facing our planet as we move headlong into the urban twenty-first century marked by intensifying urbanization and ever-growing reliance on urban infrastructures to provide for the essential needs of humanity.

The aim of *Disrupted Cities*, then, is to study moments of stasis and disrupted flow as a powerful means of revealing the politics of the normal circulations of globalizing urban life which tend to fall off the radar screen of contemporary political and social-scientific debates. The book is motivated by a paradox: Studying moments when infrastructures cease to work as they normally do is perhaps the most powerful way of really penetrating and problematizing those very normalities of flow and circulation to an extent where they can be subjected to critical scrutiny.

In fact, infrastructural disruptions provide important heuristic devices or learning opportunities through which critical social science can excavate the politics of urban life, technology, or infrastructure in ways that are rarely possible when such systems are functioning normally. Disruptions and breakdowns in normal geographies of circulation allow us to excavate the usually hidden politics of flow and connection, of mobility and immobility, within contemporary societies. Occasions of immobility and interrupted flow help to reveal urban infrastructure systems to be much more than the technocratic engineer's stuff configured in value-free ways to serve some notional public good often imagined.[3] Instead, they emerge, fleetingly, as materializations of the starkly contested and divided political, ecological, and social processes which tend both to characterize contemporary cities and to shape the configurations of the flows, and immobilities, that sustain global capitalism. Studying infrastructural disruptions critically thus allows us to do much more than learn policy or planning lessons about how to avoid repetitions of such events or how to ameliorate their effects. It also brings major opportunities to rethink and retheorize the nature of contemporary urban life.[4]

Before introducing the chapters of the book in more detail, it is necessary to understand our starting paradox of the book in more detail. To do this it is necessary to develop an introductory exploration of the nature of urban infrastructure and its disruption.

THE LUSTER OF INFRASTRUCTURE: MASTER NARRATIVES AND THE EDIFICE COMPLEX

"Cities are the summation and densest expressions of infrastructure," write Herman and Ausubel. "Or, more accurately, a set of infrastructures, working sometimes in harmony, sometimes with frustrating discord, to provide us with shelter, contact, energy, water and means to meet other human needs."[5] By sustaining flows of water, waste, energy, information, people, commodities, and signs, massive complexes of contemporary urban infrastructure are the embodiment of Enlightenment dreams of the social control of nature through advances in technology and science. They are a prerequisite to any notion of modern "civilization." They are at the heart of the ways in which cities act as the main centers of wealth creation and capital accumulation through extending their control and appropriation of labor power and all sorts of resources over distant territories, people, and ecosystems. They have tended to become inexorably woven into notions of the modern state and modern identities associated with nationhood. And infrastructure networks are always at the center of discussions about urban futurity and the impact that new waves of technological innovation will have on our rapidly urbanizing planet.

In fact, the often deterministic and teleological tendency within Western history and archaeology to label entire social and historical "ages" as "stone," "iron," "space," "railway," "nuclear," "jet," "information," "biotechnology," or whatever else, says much about the tendencies, especially within Western culture, to reduce complex social, political, and cultural realities of entire civilizations to their purportedly dominant technological circuits. These are often portrayed unproblematically as a linear march of progress—a simple process whereby faster, more capable, and more efficient infrastructures and technologies are invented, introduced, and used as a means to deterministically reorganize economies, societies, cultures, and geographies in increasingly urban ways.

Assumptions that societies are deterministically shaped by technological systems in a simple and linear way often also underlie efforts by state policy makers or city boosters to legitimize the infrastructural megaprojects that so characterize urban politics in so many cities around the contemporary world. Symbols of urban and national arrival, power, status, and kudos, gleaming new airport terminals, city-size container ports, massive highways, huge electrical generating stations, elaborate water treatment or desalinization complexes, and the high-tech accoutrements of satellite dishes, broadband lattices, and telecom masts are often still served up as icons of brave new urban or national futures. Such projects form a major element within the broader range of grandiose development projects that national political coalitions and city governments strive to build as their legacies. This edifice complex is often preoccupied with producing symbolic architectures of hypermodern mobility and circulation as demonstrations of the "globalness," "creativity," or cultural and economic power of a city within the world economy.[6]

Both linear notions of a series of societal "ages," shaped deterministic by a march of successive infrastructural innovations, and the tendency to fetishize new pieces of gleaming infrastructure, work to obfuscate the continued importance of less glamorous and long-standing infrastructural circuits of the city. Such thinking thus works to deny the ways in which multiple infrastructure systems tend to intimately rely on each other in producing the wider circulations and metabolisms of the city.

Rather than being viewed as separate infrastructural circuits which somehow work to supersede each other along the march of some putative technological progress or modernization, it is much more useful to explore how water, sewerage, waste, transport, energy, and electronic communications systems evolve and relate together.[7] An excellent case study is provided here by the prosaic example of the latest styles of fuel pumps common at U.S. gas stations. Such pumps allow users to insert a credit or debit card directly into the machine to pay for the transaction. The pumps are, in effect, machines that seamlessly integrate the operations of (at least) four separate infrastructure systems which are strung out from the station across the wider city and, ultimately, through transnational space. These include the highway/automobile system, the computer and telecommunications system, the banks' financial and credit card and ATM checking system, and the global oil production and supply system.

The connection between electricity networks and the Internet is another powerful example of the complex interconnections which fuse different infrastructures into effectively inseparable wholes. For without the "dirty" world of fossil-fuel extraction, refinement, transport, and use, the supposedly virtual worlds of new media—the hail of electrons and pixels on screens—would instantly cease to exist. The Internet, covertly, is one of the fastest-growing polluters on the planet. It is increasingly clear that "the web is no ethereal store of ideas, shimmering over our heads like the aurora borealis," as Ginger Strand writes. "It is a new heavy industry, an energy glutton that is only growing hungrier."[8]

It is startling that, unlike new airport runways, highways, nuclear power stations, or highways, the fast-growing assemblage of new media infrastructures largely escapes the ire of environmental and climate change protestors. Why should this be so? This happens because the entire, glitzy and apparently ultramodern world of new media has been ceaselessly invoked as a means of *reducing* transport demand through some wholesale dematerialization of urban society. Even though growing Internet use patently fails to reduce aggregate physical transport flows—in fact the reverse is true—digital media use continues to have an aura of transcendence, as though the "virtual" world exists in a completely separate sphere from the messy materialities of the "real" one.[9]

In practice, utopian dreams of a wholesale virtual dematerialization of urban civilization via global circuits of new media, whilst less powerful than they were, still work to deny the materiality of cyberspace infrastructures themselves, and their absolute reliance on other less glamorous infrastructure systems—most notably, huge systems for the generation and distribution of electric power. Take Michael Benedikt's influential offering, written way back in 1991. He famously predicted that the rapid growth of digital infrastructures and interactions would work as a means of wholesale societal redemption. They would work by:

> decontaminating the natural and urban landscapes, redeeming them, saving them from the chain-dragging bulldozers of the paper industry, from the diesel smoke of courier and post-office trucks, from jet fuel flames and clogged airports, from billboards, trashy and pretentious architecture, hour-long freeway communities, ticket lines, choked subways…from all the inefficiencies, pollution (chemical and informational), and corruptions attendant to moving information attached to things across, over and under the vast and bumpy surface of the earth rather than letting it fly free in the soft hail of electrons that is cyberspace. [10]

As long as even vestigial traces of such naïve utopianism attach themselves to the new media economy and culture, it will be difficult to see the world's new media economy and fast-growing digital circuits—centered on the world's high-tech cities—as more than what Brian Carroll has called "an incomprehensible, immaterial, and abstract entity."[11]

As Carroll suggests, beyond the utopian greenwash, it is necessary to see new media infrastructures as being joined intimately to vast architectures of electricity generation, distribution, and use. "The computer tool is housed by an electrical building connected to the electrical power system," he writes. "Together this infrastructure materially represents and sustains the *trompe l'œil* of other-worldly immateriality while simultaneously depending upon a physical assemblage of wires, plugs and sockets to distribution lines and poles, transformers, transmission towers and electrical power plants. Without these extensions, Cyberspace would cease to exist."[12]

A telling example here is the stealthy proliferation of squat, anonymous business buildings, housing the Internet's power-hungry data and Web server "farms," across the peripheries and exurbs of the world's major business cities. These are proliferating so rapidly that by 2008 they consumed more power in the United States than all of the televisions in the nation.[13] Between 2008 and 2011, moreover, this electricity use is expected to double.

Google's recently built server farms on the banks of the Columbia River in the state of Oregon are an excellent example here. These have been specifically located to benefit from huge amounts of cheap power necessary to keep them and their air conditioning running (for every watt of electricity used for data processing, such servers require half a watt in cooling). This server center alone—one of dozens in Google's booming global network—requires 103 megawatts of electricity—enough to power a city of 82,000 homes. With Yahoo, Microsoft, and Ask.Com also building major data centers on the Columbia River, major new electrical generating stations are being built to cope with demand—at great public expense.

"THE FORGOTTEN, THE BACKGROUND, THE FROZEN IN PLACE"?

Whilst public commentary may celebrate certain infrastructural edifices as glamorous and worthy of (at least temporary) attention, this process often works to render the remainder of a city's infrastructural circuits as curiously invisible and mundane—even boring. Indeed, when infrastructure networks work best, and succeed in reaching mass adoption as the basis for styles of urban life, they tend to become progressively both more "ordinary" and less noticed.[14] Western urban culture, certainly, has often displayed a tendency for the technological circuits of cities to be rendered culturally more invisible—at least to powerful or hegemonic users—as their use has becomes progressively normalized.

Invisibility and Infrastructure

When the very rendering of infrastructural services as virtually ubiquitous and utterly ordinary takes place it means that their use can become taken for granted and normalized as an essential, but largely invisible, support to modern urban life. In the language of the sociology of technology, such infrastructures thus become "black boxed"[15] by their users who often have no other functional alternative to relying on the networked infrastructure systems whether it is water, sewerage, electricity, the telephone, or the automobile system. Mobile and land-line telephones, electric plug sockets, water taps, flushing toilets, Internet computers, and cars thus tend to be so utterly ubiquitous in advanced industrial societies that these apparently banal artifacts give no hint to the average user of the huge and geographically stretched infrastructural complexes that invisibly sustain them.

Bruce Mau has argued that "the secret ambition of design is to become invisible, to be taken up into a culture, absorbed into the background." He argues, in fact, that "the highest order of success in design is to achieve ubiquity, to become banal."[16] Taking such a perspective further, Mau reflects that the "automobile, the freeway, the air-plane, the cell phone, the air conditioner, the high-rise—all invented and developed first in the West, but fully adopted and embraced the world over—have achieved design nirvana. They are no longer considered unnatural. They are boring, even tedious." Most of the time, Mau argues that "we live our lives within these invisible systems, blissfully unaware of the artificial life, the intensely developed infrastructures, that support them."[17]

Anthropologists Geoffrey Bowker and Susan Leigh Star, meanwhile, suggest that "good, usable [infrastructure] systems, disappear almost by definition. The easier they are to use the harder they are to see. As well, most of the time, the bigger they are, the harder they are to see."[18] Within social scientific writing about cities, especially, the vast infrastructural circuits of the city have often emerged as little more than "the forgotten, the background, the frozen in place"[19]—a merely technical backdrop that is the preserve of engineers only. Geoffrey Bowker and Susan Leigh Star offer the banal and often universal experience of uninterrupted electricity services to power a simple reading light as an example of how infrastructures have a tendency to become taken for granted. "Unless we are electricians or building inspectors," they write, "we rarely think about the myriad of database, standards, and inspection manuals subtending our reading lamps, much less about the politics of the electric grid that they tap into."[20]

When anthropologists or sociologists define the term *infrastructure,* the ways in which it sometimes attains cultural invisibility over time is one of the key criteria that they settle on. For Bowker and Star, for example, one of the eight defining characteristics of technological systems that achieve the cultural status of "infrastructure" is that they "become visible upon breakdown." They write that "the normally invisible quality of working infrastructure becomes visible when it breaks: the server is down, the bridge washes out, there is a power blackout. Even when there are backup mechanisms and procedures, their existence highlights the now visible infrastructure."[21]

When infrastructure achieves the status of a black box, few modern urbanites venture to understand the inner workings of the technology or the giant lattices of connection and flow that link these network access points seamlessly to distant elsewheres. How many of the world's burgeoning billions of urbanites, after all, routinely consider the extraordinary assemblages of fuel sources, generating stations, transmission wires, and transformers that push electrons through the myriad electrical artifacts of contemporary urban life? Or the mass of servers, satellites, glass fibers, routers—and, indeed, electrical systems—that bring our "virtual" worlds of play, socializing, e-commerce, or communication into being? Or the globe-straddling systems of communication, data processing, financial transactions, or risk profiling that bring airliners into the sky? Or the vast subterranean worlds that bring the fresh water to the tap or faucet or remove the human waste from the toilet once flushed? Or the global supply chains that populate a supermarket shelves with produce or fill the gas or petrol station with the hydrocarbon products of the decayed forms of billions of ancient life-forms?[22]

Beyond the (usually temporary) celebration of new airports, highways, glitzy fast-rail stations, TGV networks, or broadband telecommunications, the more prosaic and banal underpinnings of modern urban life tend to populate a kind of cultural substrate to the city. Many urban networks in the contemporary city remain "largely opaque, invisible, disappearing underground, locked into pipes, cables, conduits, tubes, passages and electronic waves."[23] Once initially completed and universalized—at least in many Western cities—the water, sewerage, and electricity systems of the city tended to "became buried underground, invisible, banalised, and relegated to an apparently marginal, subterranean urban world."[24] Often burrowing underground into the dark, dirty, and dusty worlds of the subterranean city, such architectures and technologies increasingly become cordoned-off physically as well as imaginatively, abandoned to the engineers, urban explorers, or cultural marginals to inhabit, reveal, or celebrate the dark labyrinths of the subway, the sewer, the security tunnel, or the historical legacies of earlier systems of movement and mobility abandoned or forgotten below ground.[25] It is not surprising that countercultural movements as well as science fiction and urban dystopias routinely delve—quite literally—into the forgotten or abandoned subterranean circuits through which the technologies burrow and connect.[26]

The black boxing of infrastructural systems, and the failure of their users to see beyond the flowing tap, the car ignition, the computer screen, the telephone handset, or the hot stove, to the empire of functions behind the working service, has further important implications for the imaginations of urban infrastructure. Cultures of normalized and taken-for-granted infrastructure use sustain widespread assumptions that urban infrastructure is somehow a material and utterly fixed assemblage of hard technologies embedded stably in place, and which is characterized by perfect order, completeness, immanence, and internal homogeneity rather than leaky, partial, and heterogeneous entities.

The combination of the social normalization of uninterrupted and ubiquitous service and the cultural invisibility of so many of the subterranean artifacts which bring infrastructural services into being, mean that entire technological systems can effectively become black boxed from the point of view of their millions of users. At the same time, our tendency to take infrastructural complexes for granted still means that, for many Western urbanites, at any rate, "technical systems [now] conjure up images of stability and permanence."[27] This is because of their historical evolution from small, fragmented, specialized systems to integrated, often (quasi) universal, and technologically standardized ones that can be regarded as "functional subsystems of society as a whole."[28]

Because of the apparent permanence of black boxed infrastructural complexes, infrastructure networks thus retain powerful images of stability. Often, they are regarded as "symbols of the complexity, ubiquity and the embodied power of modern technology."[29] This explains why urban studies, for example, still often uses a language such as "public infrastructure" or "public works," that traps these networks within a historically specific period, and consequently utterly fails to acknowledge radical current shifts in the social organization of the sectors. Urban studies appears to have difficulties acknowledging the intrinsically dynamic nature of network changes. It, too, has, in effect, tended to black box networks like electricity as permanent, ubiquitous, and banal underpinnings to urban life that don't really warrant contemporary attention.

There are dangers of overgeneralization and ethnocentrism in assuming that infrastructures always become black boxed and taken for granted, however (McFarlane, this volume). It is therefore important to stress some crucial caveats. The degree to which infrastructural services are ever normalized or taken for granted, for example, has been shown to vary dramatically both between infrastructure sectors as well as between places.[30] Not surprisingly, infrastructures tend to become normalized amongst the powerful groups and users whose interests they tend to serve. For marginal users facing poverty, even within relatively affluent cities, even relatively basic services can seem extremely precarious because of the dangers of disconnection as a result of nonpayment.[31] It is crucial to stress that, beyond the discourses of the powerful, and the comments of global design gurus like Bruce Mau, infrastructure services have "always been foregrounded in the lives of more precarious social groups—i.e. those with reduced access or without access or who have been disconnected, as a result either of socio-spatial differentiation strategies or infrastructure crises or collapse."[32]

It is equally important that Western histories whereby relatively standardized and ubiquitous infrastructure grids have been more or less constructed are not generalized as an inevitable trajectory or history for all cities everywhere.[33] In many Global South cities, for example, access to networked infrastructures has always been highly fragmented, highly unreliable, and problematic. Very often, this has applied to wealthy and powerful groups and places as well as to poor and marginal ones. (Indeed, this fact often leads city boosters to invoke major infrastructural edifices and massive demolitions of informal settlements as they strive to overcome continuous infrastructural disruptions and so become "more global" or "the next Shanghai.")[34]

This means that, in many cities, infrastructural circulations are *not* rendered as mere "technical" issues which merge into the urban background. Far from it: their politics dominates urban life and urban political discourse to a powerful extent. In such cities, moreover, improvisational capabilities to deploy backup facilities (generators, boreholes, mobile phones, chemical toilets) or coping strategies when main networks collapse tend to be well developed because the removal of such services is such a normal part of everyday life (McFarlane, this volume). Following a recent collection of papers on infrastructural politics, for example, Colin McFarlane and Jonathan Rutherford conclude that "it is clear that from the viewpoint of Bombay or Ulan-Ude [in Russia] infrastructures have rarely, if ever, been concealed or technical issues."[35]

INFRASTRUCTURE AS PRECARIOUS ACHIEVEMENT

Such caveats help hammer home a crucial point for this book: That infrastructure networks, despite their occasional veneer of permanence, stability, and ubiquity, are never structures that are given in the order of things. Instead of being static material or technical artifacts to be relied on without much thought, infrastructure networks are, in effect, *processes* that have to be worked toward. The dynamic achievement of a functioning energy, communications, water, or transport network requires constant effort to maintain the functioning system. It is easy to overplay the degree to which infrastructure networks necessarily "mature" to become socially ignored and embedded. "We sometimes seem to view mature [infrastructure networks] as invulnerable," writes sociologist Jane Summerton, "embodying more and more power over time and developing along a path whose basic direction is as foreseeable as it is impossible to detour [But] systems are more vulnerable, less stable and less predictable in their various phases than most of us tend to think."[36]

When the diverse elements are coupled and interact according to their assigned roles within any given infrastructural networks—allowing the intended effects to be expected with high reliability—sociologists of technology describe the network as stable and closed. Take the large technical system of the automobile and its related highways and service infrastructure, for example:

the techno-structure of automobile traffic is a striking example of this stability: the strongly-knit relations between automobile manufacturers and suppliers, the close intertwining of transport and taxation policy, the long-lasting tradition of motorcar engineering and the mass myth and mass practice of automobilism. Each of these relations guarantees the continuation of a technological trajectory, although the automobile traffic system has been deeply shaken by the crisis of oil supply, air pollution, and urban traffic jams. This close coupling of things, people, and signs and its continuous production by routines are the social base of the technological momentum and the myth of technics-out-of control.[37]

Despite such occasional veneers of permanence, closure, and stability, infrastructure networks are always precarious achievements. The links between nodes do not last by themselves; they need constant support and maintenance. For many of the world's urbanites, indeed—especially those in the burgeoning informal settlements which dominate many cities in the Global South—achieving an electricity, water, or communication service is the result of a constant process of improvisation (McFarlane, this volume). For such urbanites, infrastructure networks are far from being black boxes that almost miraculously and invisibly bring electricity, Internet connections, water, or food to any point or space. Instead, they are highly politicized assemblages of artifacts and practices within which continuous efforts at agency, or resistance, may—just may—allow services to be improvised, often beyond the bounds of markets and strict legality.

But even in cities where infrastructure services are associated with a veneer of stability or permanence, a vast and hidden economy of repair and maintenance is continually at work to allow infrastructural circuits to actually work.[38] Constituting at least 10 percent of most urban economies, this economy of repair and improvisation is almost invisible within the debates of urban studies. The sheer amount of economic activity generated by repair and maintenance is notable, even though it is almost completely ignored in accounts of the economies of contemporary cities. In the United States, for example, there were fully 5.82 million people engaged in installation, maintenance and repair (IMR) occupations in 2000. This figure was expected to rise to 6.48 million by 2010, a growth rate of 11.4 percent. These jobs constituted 4 percent of all jobs in the United States, making the sector one of the six most important service industry occupational groups.[39] "Think only of some of the familiar sounds of the city as an instance," write Stephen Graham and Nigel Thrift, "from the sirens denoting [automobile] accidents, to the noises of pneumatic drills denoting the constant upkeep of the roads, through the echoing clanks and hisses of the tire and clutch replacement workshop, denoting the constant work needed just to keep cars going."[40]

BETWEEN NATURE, CULTURE, SOCIETY, TECHNOLOGY, AND POLITICS: THE "INFRASTRUCTURAL TURN"

Recent research on urban infrastructure networks further underlines their precarious and vulnerable nature and helps undermine assumptions of their necessary stability or permanence. This work is the result of a remarkable growth in critical social scientific discussions about urban infrastructure, of which this book is one part. Indeed, it is now possible to speak of an "infrastructural turn"[41] in recent social scientific discussion about cities, which has paralleled the better known, but closely related, mobilities turn.[42] Work within the urban infrastructural turn has emphasized three key points.

Infrastructure as Assemblage

First, as we have already stressed, urban infrastructures are much more than a collection of technical "things" working collectively. Instead, recent work has demonstrated convincingly, inspired largely by the work of Bruno Latour and actor-network theory,[43] that it is best to consider urban infrastructures as complex assemblages that bring all manner of human, non-human, and natural agents into a multitude of continuous liaisons across geographic space.[44]

An excellent example of such a perspective comes from Jane Bennett's discussion of the U.S. electrical power grid as such an assemblage. To Bennett, the electricity grid is best understood as "a material cluster of charged parts that have indeed affiliated, remaining in sufficient proximity and coordination to function as a (flowing) system." Burning hydrocarbons, falling rain, rivers spinning turbines in dams, atmospheric weather systems, turning wind turbines, transformers, wires, plugs sockets, microprocessors, electrical consumer appliances—all relate together and gain agency alongside a myriad of social and cultural practices. "The electrical grid," Bennett writes, "is a volatile mix of coal, sweat, electromagnetic fields, computer programs, electron streams, profit motives, heat, lifestyles, nuclear fuel, plastic, fantasies of mastery, static, legislation, water, economic theory, wire, and wood—to name just some of the actants."[45]

Such a perspective underlines that any coherence that the electrical assemblage achieves as an infrastructure must never be assumed or taken as permanent and inviolable. Rather, as was made clear in the 2003 blackout, the grid is always a precarious achievement ready to untangle at a moment's notice through a myriad of possible causes (Luke, this volume). The electrical grid thus "endures alongside energies and factions that fly out from it and disturb it from within.... While they include humans and their constructions, [they] also include some very active and powerful nonhumans: electrons, trees, wind, electromagnetic fields."[46]

Cyborg Cities: Technological Mediations of Society, Nature, Culture

Second, and relatedly, recent work on the political nature of ecology has demonstrated convincingly that infrastructure networks, as well as blending the social and technical, also operate, in effect, to continually transform the natural into the cultural.[47] "Technological networks (water, gas, electricity, information etc.) are constitutive parts of the urban," write Maria Kaika and Eric Swyngedouw. "They are mediators through which the perpetual process of transformation of Nature into City takes place."[48] Thus, water systems draw water from distant biospheres, aquifers, and hydrological systems and bring this into the metabolism of urban life which in turn supports the complex modern hydrological cultures of the body and the modern city.[49] Sewerage and solid waste infrastructures radically shape urban biospheres whilst metabolizing waste matter from urban bodies, systems, or industries (Marvin and Medd, this volume). Energy systems bring fossil fuel, nuclear, hydro, wind, and solar energies into the city, which in turn sustain the complex electrical cultures of modern urban life. Urban car cultures, meanwhile, are sustained by violent conflicts over distant fossil-fuel reserves and transnational pipeline systems, and in turn play a major role in generating greenhouse gas emissions which tend to impact most negatively on far-off people and places through their contribution to sea-level rises and climate change.

Such perspectives demonstrate that infrastructural assemblages are involved in the active social production of urban natures, for example, when the hydrological systems of entire continents are shaped over centuries by practices of urban water engineering and river management, or through the ways in which capitalism sustains long-distance resource-grabs—for food, energy, or water—that add to the power of political or economic urban elites. Such productions of nature are profoundly political, even though these politics are often obfuscated by conventional ways of thinking about infrastructures as being wholly technical, separated from the entirely different and equally asocial domain of "nature."[50]

Crucially, such a perspective underlines the cyborg nature of contemporary urbanization: the ways in which the technological circulations sustained by infrastructural assemblages inseparably blend together the social relations of urban life and the relations between cities with the natural and biospheric processes upon which they rely.[51] Thus far, discussions of the blending of the biological, technological and the social into cybernetic organisms or cyborgs have most often focused either on sci-fi future dystopias or utopias; on the intimate blending of the human and technological within military or space programs; or on experiments within which bodies are artificially engineered or individual human bodies are blended seamlessly with individual technical or electronic systems or prosthetics.[52] However, Matthew Gandy and Antoine Picon,[53] amongst others, have suggested that cities and urbanization can also be viewed profitably as cyborg processes themselves.

The advantage of this approach is that it works to destabilize conventional notions that technologies somehow impact autonomously on the social world. In asserting that the cyborg materiality of a city's infrastructural circuits necessarily blends bodies, technologies, social relations, and biospheric systems, cyborgian ideas of the city also help us to assert that notions of urban politics must now move far beyond the conventional physical limits of the city. Matthew Gandy, for example, asserts that the "modern urban home, in particular, has become a complex exoskeleton for the human body with its provision of water, warmth, light, and other essential needs. The home can be viewed as prosthesis and prophylactic in which modernist distinctions between nature and culture, and between the organic and the inorganic, become blurred."[54]

Political Infrastructures

> Study a city and neglect its sewers and power supplies (as many have), and you miss essential aspects of distributional justice and planning power.[55]

Finally, and again relatedly, the recent profusion of critical research on infrastructure networks has sought to overcome the widespread tendency within popular discourse and social science alike to cast these systems as apolitical, "boring," or merely "technical" domains that can be satisfactorily partitioned off within the worlds of specialist engineers. A starting point for this work has been the criticism that, in both popular and academic discourses about cities and urban life, infrastructure networks have "too often [been] relegated infrastructures to an apolitical context or backdrop, as not worthy of attention, too hidden from view."[56]

In order to assert the highly politicized nature of infrastructure it is necessary to get to grips with what Susan Leigh Star has called the "hidden mechanisms subtending those processes more familiar with social scientists."[57] A key starting point here is that the construction of spaces of mobility and flow for some always involves the construction of barriers for others.[58] The construction, maintenance, and operations of a transport, water, energy, or communications grid tends to privilege certain more powerful spaces and users over others. The material assemblages involved have to be immobilized in geographic space in order to facilitate mobilities and circulations for selected users populations and economic groups.[59]

Infrastructure networks are thus thoroughly political constructions which tend to embody "congealed social interests."[60] This perspective emerges more easily when they are seen as messy assemblages or cyborgian complexes rather than merely the "technical" domain of engineers. (A key theme in cyborg science fiction, of course, has been the vulnerability of urban life to an increasingly automated, "intelligent" or biologically engineered technics out of control.)[61]

"For the person in the wheelchair, the stairs and door jamb in front of a building are not seamless subtenders of use, but barriers. One person's infrastructure is another's difficulty."[62] Indeed, social biases have always been designed into urban infrastructure systems and their abilities to respond to crises, collapse, or disruption, whether intentionally or unintentionally. However, the luster of highly symbolic infrastructure megaprojects, in particular, tends to obfuscate this point. Thus, celebrations about the ways in which fast-rail networks speed up the circulation of people within Europe need to be tempered by attention to the ways that they can actually make intervening spaces less accessible from city cores because local train services are often sacrificed as a result of their development.

Similar logics of bypass are evident in the pipelines of potable water which thread across the surface of Mumbai, lacing together the gated communities of the affluent, whilst providing no access to the informal cities which they bisect. Indeed, tendencies to glorify new airport extensions or highway networks in Global South megacities need to be confronted with the way informal districts are often completely erased to make way for such projects. Finally, suggestions that the opportunities of people are always improved by access to Internet computers need to be tempered by research into the ways in which software code is being used by corporate service providers to automatically sort, prioritize, and even dump electronic traffic within Internet, call-center, and e-commerce systems, as they are organized to allow privileged groups to bypass the congestion caused by the sheer weight of traffic and so enjoy a premium service.[63]

On other occasions the negative effects of infrastructural construction and use are rendered invisible by geographic or temporal distance between use and effect. The costs and risks associated with the construction of a global airport system for the "kinetic elites," who most benefit from, or control, corporate globalization, tend to fall on poorer groups, "kinetic underclasses,"[64] or distant countries facing the most cataclysmic and immediate consequences of airport expansion or climate change. Discussions about "peak oil" and the dependence of urban energy and transport infrastructures on fossil fuels, for example, rarely encompass the catastrophic impacts that climate change and rising sea-levels are already having on many marginal and peripheral people and spaces in the contemporary world.[65] The violence, destabilization, and insecurity experienced by many oil-rich regions, as imperial scrambles and wars center on appropriating the world's remaining fossil fuel reserves to feed burgeoning, largely urban demands, are also apposite here.[66] Similar quasi-imperial scrambles to buy up prime agricultural land in the Global South, to feed either the human mouths of Northern cities in the future,[67] or, through biofuel production, the growing populations of automobiles, are crucial to any discussions about ecological "sustainability" or food security in a world of rapid urbanization.[68]

With global trends toward privatization and liberalization over the past three decades, many infrastructures built within the contexts of public or private monopolies and aspirations toward universal services for all now operate according to imperatives of profit maximization and the prioritization of privileged users and markets. This "splintering" of infrastructures can involve the construction of new "premium" spaces or networks such as TGV lines, electronically tolled highways, skywalk city streets, privatized streets, or broadband communications networks, which bypass or become removed from the legacies of the inherited infrastructures.[69] More subtly, it may encompass the withdrawal of essential services from poorer or less profitable groups or spaces, as efforts concentrate on addressing the more profitable market segments.[70] It may be associated with tendencies to privilege selected premium infrastructures whilst reducing essential public investment and maintenance from the wider inheritances of infrastructure converging entire cities, regions, or nations.

Finally, the resilience of infrastructures may be severely compromised as they are actively reorganized to maximize profit or return or absorbed wholesale within predatory models of neoliberal financial capital. The short-termism, predatory tendencies, and risk of asset stripping which attend the progressive merging of infrastructural and finance capital in many cases have been a notable element of the current collapse of neoliberal financial capitalism.[71] Often, indeed, speculative finance capital has not only preyed on, and profited from, crises and disruption, it has constituted political economic conditions necessary to *generate* such crises, whether they be wars, postwar reconstruction efforts, financial meltdowns, or predatory disaster related operations.[72]

The spectacular financial collapse of 2008 and 2009, of course, has brought its own debilitating disruptions as bankruptcies and frozen credit markets have left a multitude of projects to build, ameliorate, or improve infrastructural networks canceled or on ice. In the United Kingdom, for example, the construction of new schools is now almost paralyzed because it relies so heavily on neoliberal development models based entirely on private finance capital.

Sparked by the collapse of California's deregulated electricity system in 2001, Gene Rochlin argues that this example, along with other notable cases such as the notoriously disastrous privatization of the UK rail system, or the often chaotic trajectories of U.S. airline or phone systems, demonstrates that regulatory shifts in infrastructure development, legitimized by the dogmatic application of neoliberal economic dogma, can be as disruptive to reliable services as any natural disaster or terrorist attack.[73] In many cases, at least in the Global North, he suggests that liberalized infrastructure models have in many cases involved an abandonment of social regulations sustaining progressive cross-subsidies and universal distribution of service built up through Keynesian economic regulation and the development of welfare states.

Whilst they were far from perfect and extremely varied, Rochlin suggests that the widespread replacement of socialized models of long-term, Keynesian infrastructure regulation with models privileging short-term or speculative profit, the role of finance capital, and individualized notions of consumption, have actually often worked to increase risks of failure and disruption in many cases. The shift from relatively coherent and equitable assemblages, based on ideas that infrastructures are economic "natural monopolies," to a baroque complexity of fragmented institutional providers, certainly seems to underline that many infrastructure regulators do not seem to understand that complex infrastructural assemblages are not simple commodities distributed to consumers within mythically perfect private markets. In the California electricity and UK rail cases, in particular, Rochlin argues that the result has been a startling increase in user prices, a rapid growth of disruptions, vulnerability and volatility, an overreliance on the short-term volatilities of financial markets, and a growth of predatory corporate behavior where operators work to amass their own quasi-monopolistic powers whilst sucking in more and more hidden subsidy from governments.[74]

Jane Bennett, though, reminds us that the technical elements of infrastructural assemblages have their own agency within infrastructural disruptions, and that it is inadequate to merely read off the frequency or nature of disruptive events from political economic or regulatory transformations. This makes the postmortems that follow major collapses like the 2003 U.S. blackout especially interesting (see Luke, this volume). "A distributive notion of agency does interfere with the project of blaming," Bennett writes. "But it does not thereby abandon the project of identifying…the sources of harmful effects. To the contrary, such a notion broadens the range of places to look for sources. "[75] In this case, Bennett urges us to look to "selfish intentions and energy policy that provides lucrative opportunities for energy trading while generating a tragedy of the commons"; at "the stubborn directionality of a high-consumption social infrastructure"; and at "the unstable power of electron flows, wildfires, ex-urban housing pressures, and the assemblages they form."[76]

Within such a context, the degree to which infrastructures are "resilient" to all manner of disruption is both socially constructed and politically contested.[77] Infrastructures, and the way they are maintained (or not maintained) through continuous, often hidden, work, distribute risks unevenly. Despite the proliferation of state "critical infrastructure" policies since September 11, 2001,[78] extremely uneven resources and effort are put into establishing back-up systems or alternative sources of supply for different portions of the population served by a system. As Julian Reid has argued, these policies reveal much about the way Western political thought stresses infrastructural circulation as the very means of life—the key biopolitical basis of political power. "The defence of critical infrastructure," he writes, "is not about the mundane protection of human life from the risk of violent death at the hands of terrorists." He sees it, rather, as:

> a more profound defence of the combined physical and technological infrastructures on which global liberal regimes have come to depend for their sustenance and development in recent years. "Quality of life" is deemed inextricably dependent in these documents on the existence of critical infrastructures. Terrorism is a threat to these regimes precisely because it targets the critical infrastructures which enable the liberality of their way of life rather than simply the human beings which inhabit them.[79]

Critical infrastructure response plans, moreover, are often extremely biased toward the needs of hegemonic elites or capital groups. Crises often bring unerring focus on such glaring biases. Eric Klinenberg's ground-breaking research on the infrastructural and institutional response to Chicago's devastating 1995 heat wave, for example—which killed over 700 people—demonstrated systematic failures to address the extreme vulnerabilities of the city's poorest African-American neighborhoods.[80]

Indeed, in many cities around the world, instead of addressing the needs of vulnerable groups and communities, major state and corporate investments go to sustaining continuities of flow and circulation, by providing backups to key electrical, communication, data processing, transport, or other infrastructures that sustain the main nodes and enclaves of the globalized corporate economy. (The geographical organization of corporate systems based on the seamless digital and material interlinking of globally stretched nodes with their just-in-time logistics flows only adds to the imperatives here).[81] Indeed, a key part of the spatial production of extrajudicial or extraterritorial "free trade" zones, "export processing" zones, or "special economic zones" that are so central to global economic geographies these days, is the installation of systems that secure the circulations of capital even during times of crisis, whilst undermining labor and environmental controls at the same time (Cowan, this volume).[82] Just beyond the borders of such nodes, however, service interruptions and disruptions often expose a stark lack of alternatives.

The political nature of infrastructure disruption is often rendered invisible by media discussions of such events as mere "technical" malfunctions or environmental "Acts of God." The notion that urban natures are actually produced through the long-term agency of political infrastructural assemblages renders such perspectives unhelpful, however. Such understandings hammer home the key point that, in infrastructurally mediated natures, there is no such thing as a natural disaster.[83] Such problematizations rarely encroach on popular discussions, however. Coverage of the impact of Hurricane Katrina on New Orleans in 2005 (see Sims, this volume), for example, often failed to attend to the ways in which "nature" in this case was actively produced by the long-term engineering of the mouth of the Mississippi by the U.S. Army Corps of Engineers. Over centuries, the flood protection and river engineering systems in and around the mouth of the Mississippi worked to produce highly uneven geographies of exposure to flood risk, which were racially biased in ways that were only fully exposed by the disaster itself.[84]

SECURITY AND INFRASTRUCTURE: NORMALITY AND EVENT

It is important to stress how the politics of international, national, state, corporate, or urban security are now especially preoccupied with the sense that the infrastructures sustaining urban life provide an urban Achilles heel to be attacked and exploited by all manner of state or nonstate threats (Graham, this volume). John Robb, an influential U.S. commentator on the links between infrastructure and security, regularly argues that "global infrastructure networks are the Achilles heel of the great powers. They form the basis of our [sic] wealth and our daily function and they remain extremely vulnerable."[85] Some social theorists argue that new styles of state war and political violence are emerging based on manipulations of the specter of such threats.

Allen Feldman, for example, has noted the proliferation of open-ended and deterritorialized "securocratic wars" such as the recent U.S.-sponsored transnational "wars" on drugs, crime, terror, illegal immigration, malign software code, and biological threats. These are organised around purported threats of urban infrastructure disruptions, by the notion that good and virtuous infrastructural mobilities—sustaining urban consumption, corporate globalisation, legitimate tourism etc.—are continually threatened by the colonization of all manner of malign circulations and flows which are deemed to contaminate societies and threaten the social order internally and externally simultaneously. Such securocratic wars also tend to fetishise certain risks—the risks of terrorist attack, for example—whilst neglecting often much more widespread ones (such as the global plague of over 1.2 million automobile deaths per year, or the risk of coastal flooding of major cities that comes with rising sea levels).[86]

Contaminating circulations that are at the center of securocratic warfare include infrastructural terrorism against transport, water, energy, digital or nutritional and agricultural infrastructures, demographic infiltration, "illegal" immigration, terroristic financial transactions, malign software code, and pathogens and disease (SARS, bird flu, swine flu, Mad Cow disease).[87] Unknown and unknowable, these varied and dispersed existential threats are deemed to lurk invisibly within the very interstices of urban and social life, blending invisibly with, and operating through, the very infrastructural assemblages of the city (on the SARS case, see Ali and Keil, this volume).

At their root, open-ended, securocratic wars are today invoked as attempts to police subnational and supranational dichotomies of safe and risky places within and beyond the territorial limits of nation-states.[88] Central here is the distinction between an event and the normal, societal background. Thus, "security events" emerge when what Feldman calls "improper or transgressive circulations"[89] from the range of putative threats become visible within and through the disruption or appropriation of urban infrastructure, and are deemed to threaten the "normal" worlds of transnational capitalism.

At the same time as transgressive circulations are rendered as security events or crises, the global logistics, commodity, tourism, labor, migration, and financial flows sustaining neoliberal capitalism are rendered invisible by their very "normality." These are the nonevents of "safe circulation" organized through infrastructural backgrounds to link up transnational archipelagos of safe or risk-free spaces for capital and socioeconomic elites. "The interruption of the moral economy of safe circulation is characterized as a dystopic 'risk event'," Feldman suggests. "Disruption of the imputed smooth functioning of the circulation apparatus in which nothing is meant to happen. 'Normalcy' is the non-event, which in effect means the proper distribution of functions, the occupation of proper differential positions, and social profiles."[90]

Crucially, however, circulations deemed malign, along with their associated humans—the electronic financial transaction, the data shadows generated by a travel reservation, the human civilian on the subway, street, or aircraft—tend to be indistinguishable from the wider urban mass. Practically, securocratic war thus must center on the construction of complex systems of "data mining" and *anticipatory* surveillance. In a very real sense, then, the proliferation of unending, unbounded, and transnational "wars" against malign circulations or disruptions within urban infrastructure systems brings a radical convergence between notions of state warfare, political violence, and security and questions of the politics of urban infrastructure. This has crucial implications for the politics of both globalization and cities, as vague "security" threats are invoked around the world to cordon off and fortress strategic urban spaces (Cowan, this volume.) The pervasive cultures of fear stirred up and manufactured by securocratic war, for example, often mean that everyday failures, accidents, or disruptions that have nothing whatever to do with "terrorism" are initially interpreted as though they are its direct results.

Paradoxically, however, the invocations of "security" that sustain both securocratic war and constructions of securitized urban spaces have little to say about the threats to security that are doing so much to undermine levels of wealth, well-being, and security in our urbanizing world: a global meltdown of neoliberal financial capitalism and global climate change driven by profligate consumption and constructions of infrastructure to sustain it. These, and surrounding discourses about risk, fear, and security, are at the heart of the debates which follow in this book's discussions about infrastructure disruptions.

UNDERSTANDING INFRASTRUCTURE DISRUPTIONS

City-dwellers are particularly at risk when their complex and sophisticated infrastructure systems are destroyed and rendered inoperable, or when they become isolated from external contacts.[91]

The continuous reliance of urban dwellers on huge and complex systems of infrastructure stretched across geography creates its inevitable vulnerabilities. When infrastructure services have become taken for granted, paradoxically, as we have seen, it is often the moment when the blackout occurs, the server is down, the subway has a strike, or the water pipe ceases to function that the dependence of cities on infrastructure networks becomes most visible. In such circumstances, "for most of us," writes Bruce Mau, "design is invisible. Until it fails."[92]

Sudden disruptions to the complex and "cyborgian" assemblages that sustain infrastructure networks brings with it the startling paradox with which we started this chapter: The unexpected *absence* of functioning infrastructure works to underline the very (albeit useless) presence of the vast stretched-out system that usually remains so invisible.[93] When they have become stable and taken for granted, interruptions in power, clean water supplies, the arrival of fresh food, the ability to move commuters, tourist or business travelers, the means of flitting electronic data, money, communication, or video around the planet at the speed of light, or the means of shifting waste and sewerage away from teeming cities, immediately work to make the vast complexes of infrastructure on which urbanites continually rely starkly visible—if only until normal services are resumed. Sociologists of technology call this a process of "unblackboxing": a social process, the opposite of the black boxing process discussed above, through which the complex system and technologies upon which everyday life relies, which are normally kept within a black box that is only penetrated by specialist engineers and policy makers, are suddenly clearly revealed.[94]

Interconnection and Cascading Effects

During infrastructural disruptions within contexts where infrastructure has become taken for granted, to adopt Erving Goffman's terms, the built environment's "backstage" becomes momentarily "frontstaged."[95] But because infrastructures that are usually considered separate are actually woven together in all sorts of mutually dependent ways[96]—as with the Internet/electricity example already discussed—disruptions to one infrastructure quickly cascade through other systems in unpredictable ways (see Little, this volume).[97] As Charles Perrow's highly influential book *Normal Accidents*[98] demonstrates, tightly interconnected infrastructures "predictably fail but in unpredictable ways."[99] Crucially from the point of view of this book, disruption or destruction at one point in a water, transport, communication, or energy grid tends to move through the whole system. And because these systems are densely interlinked and mutually dependent—or are tightly coupled in engineering parlance—disruption in one tends to cascade to others very quickly. Thus, when the baggage handling facilities at Heathrow's new terminal 5 failed to keep up with passenger throughput in March 2008, cascading effects quickly disrupted the entire planet's airline system. When thieves literally cart off electrical copper networks—affected by the high price of the metal—as has been common in China, Europe, Russia, or the Global South within the last decade, complex chains of deelectrification can quickly paralyze the multiple electrical circuits and digital or physical circulations of urban life.

Given that all of the "Big Systems" of infrastructure that sustain advanced, urban societies are profoundly electrical, city residents become, in particular, "hostages to electricity."[100] This is because it is very difficult to store large quantities of electricity. In an electrical blackout it is not just electric lighting that fails. Electrically powered water and sewerage systems tend to grind to a halt. Public transportation stops. Food processing and distribution is disabled. Health care becomes almost impossible. Even the Internet ceases to function. Large scale, cascading infrastructure failures, particularly between electricity and transport outages and other systems, demonstrate that there are many orders of cascading effects. For example, Richard Little recounts how, in May 1998, the failure of just one satellite terminated the operation of 80 percent of all U.S. pagers, disrupted ATM and credit card transactions systems, interrupted emergency health care communications systems, and brought chaos to the complex, just-in-time systems in place in health care systems.[101]

From Apocalyptic Fears to Cultures of Repair

Just as cultural debates about infrastructure tend to privilege glitzy and glamorous infrastructural edifices, cultural commentaries about infrastructural disruptions tend to be about the spectacular collapse of whole cities, societies, or civilizations, rather than the mundane interruptions and repairs and improvisations that constitute the quotidian existence of urban dwellers.[102] As we have seen, in many cases, infrastructural services do not become black boxed and taken for granted because their use is always precarious and unreliable. This point is especially important in the parts of Global South cities where rudimentary or improvised access to power, water, fuel, food, or sanitation, beyond the limits of formal economies, is all that is possible (McFarlane, this volume).

Rather than swarming masses of repair workers, or urbanites tinkering with the prosaic technicalities of urban life, or dealing with cascading infrastructural disruptions, though, we find that films, video games, and novels are endlessly preoccupied with fantasies of complete societal or urban collapse replete with annihilated cities and complete societal breakdowns.[103] We have mass disasters, wholesale loss of life, the repeated end of cities per se,[104] and reversals to preindustrial existence for small bands of hardy survivors, rather than the improvised coping strategies of users and providers in dealing with day-to-day infrastructure disruptions. We also encounter widespread social and political discussions of how past societies have collapsed[105] and how contemporary civilization is facing a future wholesale collapse characterized by resource exhaustion, runaway climate change, growing demographic pressure, and spiraling warfare, rather than the prosaic experiences of sewer overflows, transport disruptions, or energy blockades.[106] And we rehearse the millennial speculations of endless predictions of apocalypse by "cyberterror"[107] or malign software running out of control—think of the debates that surrounded the Y2K problem in the run-up to the year 2000—rather than the endless and deeply prosaic software glitches, crashes, and the continuous repair necessary to run a simple Windows PC, a city electrical system, or an organizational computer network.[108] In addressing the infrastructure disruptions in this book, therefore, we need to be especially mindful of the continuous, invisible work necessary to bring about infrastructural circulation even when infrastructural assemblages are working "normally."[109]

Of course, fear, apocalypse, and catastrophe sell; routine portrayals of prosaic improvisation don't. Disaster genres in film, video games, and novels, after all, tap into a widespread sense of apocalyptic dread about the fragility of urban life in times of growing environmental stress, burgeoning population, and the growing sense of imminent or existing resource exhaustion, catastrophic climate change, or biodiversity collapse. More generally, there is a widespread cultural sense of the flakiness of many of the essential infrastructural services organized through often unreliable "kludges" of myriads of software fudges lacing together massive computer communications systems.[110] "Fear of the dislocation of urban services on a massive scale," writes Martin Pawley, is now "endemic in the populations of all great cities," simply because contemporary urban life is so utterly dependent on a huge range of subtly interdependent and extremely fragile computerized infrastructure networks.[111]

Disruption and Digitality

Such is the cyborgian nature of computerized infrastructure systems that disruptions to normal services are sometimes moving from the status of inconveniences to that of life-threatening events. Taking an unusually reflective and critical stance for a software engineer, Bill Joy, cofounder of Sun Microsystems, caused a furor back in 2000 amongst readers of the bible of the high-tech elite, *Wired*. He suggested that the mediation of human societies by astonishingly complex computerized infrastructure systems will soon reach the stage when "people won't be able to just turn the machines off, because they will be so dependent on them that turning them off would amount to suicide."[112] Such concerns fuel an entire publishing industry emphasizing that the lurking actions of distant "cyberterrorists" could bring an "electronic Pearl Harbor" to the U.S. nation by sowing mass destruction and death by bringing paralysis to air traffic control, logistics, electricity, water, or other critical systems through the use of malign code (see Graham, this volume).[113]

Computer worms and viruses are often deemed by U.S. security commentators to be mere trial runs for such mass digital paralysis. The "I Love You" or "Love Bug" virus, launched by a college student in the Philippines on May 3, 2000, remains a powerful example here. This virus moved to infect 45 million computers in at least twenty nations across the world within three days, clogging and destroying corporate e-mail systems in its wake. Overall damage was estimated at well over $1 billion and many Fortune 500 companies were substantially affected. The virus also exposed some of the transnational tensions and inequalities that surround corporate IT. Some newspapers in the Philippines, for example, expressed national pride that the country could spawn a hacker that could bring the highly fragile computer communications systems of Northern corporations to an (albeit temporary) collapse.

On February 1, 2008, meanwhile, just off the coast of Alexandria, Egyptian fishing trawlers inadvertently severed the optic-fiber lines which work to continually bring the "virtual" interactions of global finance, global telephony, and the Internet into being. In an instant, entire portions of the planet—on this occasion, India and the Middle East—suddenly experienced "network unavailable" signs on their computers, the collapse of stock markets, or the disappearance of their telephone call tone. Again, flurries of media reports momentarily exposed the geographies and politics through which glass strands are weaved together at the bottom of the world's oceans to sustain burgeoning electronic interactions between global archipelagoes of high-tech cities and urban economies. For a day or two, the serious newspapers were full of detailed maps of the world's optic fiber networks. Once the crisis was over, these geographies sediment back into the collective unconscious until another interruption occurs.

The reorganization of every circuit and aspect of modern urban life through incredibly complex digital control systems[114] clearly adds a new and vital twist to discussions about the politics and impacts of infrastructure disruptions. Crucial within such debates is the growing awareness that, far from being rational, orderly, or even explicable, digital systems often display a kind of vitalism and nonlinear complexity that it can be difficult even for experts to explain how they work (or, equally, don't work).

Beneath the techo-boosterism of the "digital" or "networked society," then, and far removed from the rapidly receding utopianism of the likes of Michael Benedikt discussed above, the prosaic and everyday realities of using contemporary computer systems is, in many ways, constituted through continuous repair and maintenance. Ellen Ullman stresses that the Y2K "crisis," in particular, hammered home the fact that contemporary ICT systems are not "shining cities on a hill—perfect and ever new—but something more akin to an old farmhouse built bit by bit by non-union carpenters."[115] "Glitches, patches, crashes," the crisis revealed, were "as inherent to the process of creating an intelligent electronic system as is the finely tuned program, the gee-whiz pleasure of messages sent around the world at light speed."[116]

During the Y2K crisis, even computer and software engineers often had little idea of the full archaeological sedimentation of decades worth of software within their computer networks. These underlined what Ullman calls the "near immortality of computer software"—the way new software often merely aggregates around the kernels of very old systems. Resulting systems are thus inevitably going to be unreliable to an often unknown extent. In the event, only one of the largest concerted repair operations in human history, in the years leading up the turn of the millennium, was able to avert the mass failure of a whole host of transnational ICT systems, and the interdependent infrastructures that they sustain.

Large amounts of any investment of time and money in keeping an IT system running is inevitably spent confronting the need for continuous software and hardware upgrades and maintenance; installing software patches to iron out a continuous stream of identified flaws; addressing the malignant code that is continually unleashed into the world; organizing secure backup systems to maintain data in the event of a major crash; and training and equipping the staff, facilities, and services to offer such continuous repair services. To take just one example, within Metropolitan Chicago in 2003, "computer maintenance and repair" constituted 4 percent of all jobs in the city (5,679 jobs in all).[117] Moreover, a burgeoning universe of software support and call center help lines, spread right across the world to service the major markets of northern metropolitan areas, constitutes one of the world's fastest-growing industries. Here we confront transnationally configured networks, organized through computer systems, which link consumers in the Global North to advisors in the Global South, and whose very raison d'être is the continual requirement of users to deal with mass, routine failure in computer systems. Spurred on by 9/11, new urban landscapes of repair and maintenance have even started emerging around the cores of the world's great cities, as emergency computer centers, reinforced and windowless like Cold War bunkers, are built to be occupied within minutes in the event of a major disruption or crisis.

Fears of the complete collapse of digital circulations are paralleled by updated variations of long-standing fears of autonomous or cyborgian technics running amok. Events where largely autonomously software, linked into cyborgian infrastructural assemblages, automatically triggers devastating actions—such as with the shooting down of Iran Air flight 655 by the automated computer systems of the U.S. Navy's U.S.S. Vincennes—add a new twist to such risks.[118] The interdependence linking electronic communications infrastructures and other infrastructural assemblages also means that electronic disruptions and signals are likely to unpredictably disrupt flows of more prosaic and less glamorous infrastructural circuits, as they themselves become organized through networked computer systems. In February 2006, for example, the cars of drivers on a coastal road in Norfolk, England, started to mysteriously grind to a sudden halt. Local mechanics were completely flummoxed ("It's like the X-Files, isn't it?" one was reported to have said). Eventually, it became clear that the control systems of the latest computer-controlled cars were being disrupted by powerful radar signals from a nearby early-warning station (as we will see later in this book—Graham, this volume—these very means of electronically disabling modern infrastructures are at the heart of emerging military doctrine surrounding electronic pulse weapons).

Conversely, disruptions in physical transport infrastructures can have far-reaching effects on electronic communications circuits. This is because, at least within cities, these circuits are invariably layered within conduits that parallel physical roads, subway tunnels, or rail systems to minimize costs. In July 2001, for example, a fire on a train in a tunnel beneath central Baltimore brought network disruptions to e-mail and Internet traffic in places as distant as Atlanta, Seattle, Los Angeles, and even Lusaka, Zambia.[119] The fire revealed the continuing role of key U.S. metropolitan areas as the hubs of the vast majority of the world's electronic traffic—a legacy of the military and Cold War origins of the system, and the global commercial dominance of U.S. telecommunications and Internet firms. Against the rhetoric of steplike shifts toward a dematerialized information society demonstrated by writers like Michael Benedikt, discussed above, events like the Baltimore train fire hammer home, rather, that "new infrastructures do not so much supersede old ones as ride on top of them, forming physical and organizational palimpsests—telephone lines follow railway lines, and over time these pathways have not been diffused, but rather etched more deeply into the urban landscape."[120]

Any confrontation with the realities and myths surrounding digital vulnerability must start by stressing that, whilst these risks are real, debates about the lurking threats of complete societal paralysis through "cyberterror" are often both overblown and manipulated for political ends. Many argue that "cyberterror" is actually much less of a threat than is often imagined. Some suggest that such attacks would fail to generate the catastrophic spectacle at a single point in space and time that is generally the objective of terrorist arracks. Others suggest that the messy and improvised nature of infrastructural assemblages means that they are actually much less vulnerable to the infiltration of malign code than is often imagined.[121]

Indeed, sometimes a *lack* of connectivity is a cause for active celebration and publicity. The often coercive and stressful realities surrounding continuous, "always-on" digital connectivities may sometimes make a lack of functioning infrastructure attractive. The Welsh tourist board, for example, recently launched a major advertisement campaign under the strap line—playing on the UK environmental designation "Area of Outstanding Natural Beauty"—of "Area of Outstandingly Bad Mobile Reception." (Not surprisingly, however, Welsh business groups were appalled at this portrayal of the Principality as celebrating its relative technological backwardness.)

Whilst the risks of catastrophic societal collapse are certainly real, we need to be mindful of the dangers of perpetuating what Josef Konvitz—writing about the impact of Allied bombing on the infrastructures of German cities in World War II—has called "the myth of terrible vulnerability."[122] Konvitz argues that, by emphasizing the catastrophic urban impacts of disaster, collapse, or warfare, and simultaneously neglecting the way in which the often remarkable resilience of urban infrastructure combines with continuous and prosaic efforts at repair, urban literatures tend to radically overemphasize the vulnerabilities of cities to complete, sustained, and irrevocable collapse.

In the process, the continuities between the hidden and ongoing cultures of repair that characterize urban life outside of catastrophic states, and the efforts to overcome natural or human-crafted catastrophe, tend to be dramatically underplayed. I would argue, for example, that the celebrated efforts of systems engineers to bring back Web, mobile, data, financial, and TV services to the world's most connected urban place—Manhattan—after the 9/11 attacks, merely constituted concerted strategies that built on the continuous processes of management and repair that allow such usually invisible systems to exist in the first place.[123] The remarkable ways in which cities rise again after catastrophe thus have a great deal to do with the ways in which cities are being continually repaired outside more cataclysmic periods.[124]

Disruption Cultures

Recent efforts by the artistic world to represent threats of infrastructure disruption have recently received a great deal attention, most notably through the work of French theorist Paul Virilio on the links between speed, modernity, and the rapid or instantaneous diffusion of technological "accidents" through global reliance on interlinked digital control systems.[125]

Perhaps more surprisingly, popular music has also started to develop some of the most powerful representations of experiences of infrastructural disruption. Members of the Canadian band, the Arcade Fire, for example, personally experienced the huge power collapses which impacted on Eastern Canada's main cities in January 1998 because a massive ice storm led to the widespread collapse of the power transmission system.[126] Some of the 5 million people affected were without power for five weeks during one of the coldest Canadian winters). As Jacques Leslie recounts, classic cascading effects quickly brought a state of emergency to the area:

> People without power discovered just how many facets of their lives depended on electricity. Their stoves, appliances, and heating didn't work, and many telephones went out. In eastern Ontario, where 50,000 phones went dead, the electric utility, Ontario Hydro, was doubly confounded, since it depended on customers' phone calls to alert it to power failures. Throughout the affected region, all financial transactions had to be in cash, since credit card swipes and ATMs were useless. And even if drivers could find highways free of tree limbs and power lines, they could go only as far as the gas in their tanks would take them, because gasoline pumps didn't work. Most disturbing of all, at 12:20 p.m. on the 9th, the two water filtration plants that served 1.5 million people in the Montreal region went down, leaving the area with a 4- to 8-hour water supply."[127]

Arcade Fire's song, "Neighborhood #3 (Power Out)," provides a particularly visceral reflection of the experience, and hammers home the sense of modernity unraveled, lives threatened, and norms abandoned amidst darkness and cold that few had experienced before.[128]

Importantly, however, urban cultures in cities facing major infrastructure collapses are often characterized by what Michael Sorkin has called a kind of "crisis conviviality."[129] Whilst there is always a danger of romanticizing the purported return of premodern or preinfrastructural sense of community or collective action, collective improvisation often thrives within the contexts of infrastructural collapse. For example, Sorkin argues that, since the September 11, 2001 attacks, New Yorkers have deployed what he calls a new "paradigm" when responding to infrastructural crises like the 2003 blackout. In both events:

> Pedestrians took over the streets and sidewalks, not simply for walking home but as a social space. Outside every bar and bodega, a crowd gathered—it was after all almost cocktail time when the power went out—and the mood was festive. The blackout functioned as a kind of Disney version of the *blitz*. To be sure, the city was "paralyzed" but we all enjoyed the opportunity to display our civility, to use the disaster to expand the territory of public space."[130]

In the two crises in New York, such impromptu and celebratory cultures were especially energized by the way the crises brought about urban situations that progressive movements had been fighting for, in vain, for decades: car pooling, pedestrianized space, waterborne commuting, very tight controls of car use.

Meanwhile, in cities like Mumbai, where the "new normality" of continuous infrastructural disruption is a habitual element of daily urban life—especially for the poor—such disruptions are often marked by the impromptu coping strategies and altrusim of a wide range of social groups and social movements (McFarlane, this volume). Indeed, Jonathan Shapiro Anjaria underlines the dramatic contrast in the responses to the massive flooding of both New Orleans and Mumbai within a four week period of 2005. He argues that the sophisticated and collective coping strategies deployed in the latter case at the grassroots level worked to prevent the chaos and violence which so dominated the former one.[131] Moreover, the social mobilization of strategies to address infrastructure crises in many Global South cities also often works to pave the way for much larger political movements. These work to contest both the extraordinary inequalities surrounding contemporary models of urban growth in many Global South cities, and the increasing commodification of urban infrastructure (McFarlane, this volume).[132]

Disruptive Politics

Finally, it is useful to stress the growing centrality of deliberate infrastructural disruption to the political strategies of dissenters, protestors, and states and nonstate fighters alike. Public protest in cities, for example, increasingly eschews traditional mobilizations in city centers, concentrating instead on disabling or occupying the most strategically important infrastructural hubs of a city. In November and December 2008, for example, thousands of activists of Thailand's main opposition party, the People's Alliance for Democracy (PAD), occupied Bangkok's two main airports, preventing at a stroke their 125,000 daily passengers from traveling. A powerful demonstration against Prime Minister Somchai Wongsawat, whom they argued was merely a puppet of the previous ousted President, the occupation was an effective siege which completely disrupted the lucrative tourist economy of the whole of Thailand.

State military theorists are also all too aware of the debilitating cascading effects of interrupting infrastructural flows, especially electricity supplies. They have developed a range of military doctrine which legitimizes the destruction of urban societies' electrical systems as means of reducing societies deemed adversaries to a state of "strategic paralysis." The U.S. and Israeli armies, amongst others, have carefully developed a range of weapons—from giant 60-ton bulldozers equipped with claws designed to destroy roads, water systems, and power lines to "soft" bombs which rain millions of graphite coils onto electrical substations, "deelectrifying" entire adversary societies in an instant. Indeed, a complex body of military theory legitimizes such attacks as a necessary and supposedly "nonlethal" means to coerce highly urban societies by bringing about complete infrastructural disruption or devastation. However, by generating huge public health crises, as electrically powered water and sewerage systems grind to a halt and repair becomes impossible because of sanctions, such targeting actually leads to very large numbers of civilian deaths, usually amongst the old, the young, and the ill (Graham, this volume).

Nonstate fighters and insurgents, meanwhile, have moved well beyond the long-term staple of the car bomb.[133] They now attempt to appropriate airliners, subway cars, railway carriages, and buses as means, paradoxically, to devastate and interrupt the circulations of cities.[134] Infrastructures and technologies of circulation are preeminent amongst the myriad of "soft targets" that constitute contemporary cities in the eyes of such movements. They symbolize the purported arrogance of technocratic Western nations or the transnational reach and power "global" cities.[135] They provide opportunities to engineer massive media events and extraordinary levels of devastation without the use of any military weapons whatsoever. And they help generate incalculable economic costs as "normal" circulations and flows sustaining globalized capitalism are interrupted by cascading disruptions unleashed unpredictably in space and time.

In India, for example, terrorists have targeted the electrical infrastructures sustaining the country's burgeoning high-tech enclaves.[136] In Iraq, Saudi Arabia, and Nigeria's Niger Delta, meanwhile, a whole spectrum of insurgents work to destroy oil pipelines, targeting the distant supply lines of fossil fuels as a means to bring economic and political pressure on distant cities. In all these examples, perversely, "the space of the city, orchestrated by the organizational logics of infrastructure, is…precisely revealed in its destruction."[137] Political violence that is directed against infrastructure is perhaps the ultimate way to forcibly "unblackbox" infrastructures that have managed to achieve the status of perceived stability, invisibility, or permanence.

Geopolitical power, finally, increasingly centers not merely on the use and deployment of military power but on the control of the energy, water, and food resources which must continually be imported to sate the appetites of rapidly urbanizing societies. Vladimir Putin's resurgent Russia, for example, is emerging one again as a first order power largely through its continual threats and disruptions to the gas supplies that it pipes Westward toward the Ukraine and Eastern and Western Europe.

INTRODUCING THE CHAPTERS

If the city is to survive, process must have the final word. In the end the urban truth is in the flow.[138]

Disrupted Cities is made up of eight complementary chapters. Richard Little kicks off discussions by outlining the complex ways in which infrastructure disruptions tend to cascade between different networks and across geography and time in highly unpredictable and non-linear ways. Little's main concern is how public policy can possibly hope to anticipate and manage these complex cascading effects. Using a range of case studies, and the latest theoretical and analytical approaches, Little provides powerful insights into the ways U.S. policy researchers are grappling with the enormous challenges and complexities thrown down by cascading infrastructural disruption.

Benjamin Sims takes such a critical public policy perspective forward with a detailed case study of the ways in which the catastrophic effects of Hurricane Katrina on the city of New Orleans particularly affected the City's police department. How, in other words, did a public body tasked with the challenges of responding to the crisis, and reinstating their normative notion of moral order on the city, when the infrastructural assemblages upon which it itself relied were also laid waste by the crisis and its own officers were fearful and traumatized?

In chapter 4 Tim Luke deploys critical social theory to address the iconic blackout that systematically deelectrified one of the world's great metropolitan corridors—the Northeastern seaboard of the United States—on August 14, 2003. Delving deep into the politics of the event, Luke ponders on the way brittle infrastructural assemblages which are normally prone to accidents have been generalized and naturalized as inviolable bases of modernity itself within many contemporary cities.

Deb Cowen then provides an enlightening discussion of the ways in which purported threats of infrastructural disruption within globalized capitalism are being invoked by dominant power holders in the planning and development of cities with new logistical and port enclaves. Addressing the construction of an entire "Logistics City" in Dubai, and its role as a node within the U.S. Department of Homeland Security's attempt to "offshore" its port security systems across the world's main port complexes, Cowen exposes powerful intersections between urban, labor, and trade politics within the context of neoliberal and corporate globalization. She emphasizes how labor and civil rights within port cities are being systematically compromised because invocations of securocratic war are being used to securitize both global port systems and the wider urban complexes that provide the homes for major port hubs.

In contrast to the iconic and spectacular infrastructure disruptions discussed by Little, Sims, and Luke, Simon Marvin and Will Medd bring our attention to a prosaic, incremental, and hidden crisis in the urban metabolic circuits of subterranean sewerage. They demonstrate that the urban metabolisms of contemporary cities are becoming as sclerotic as are the bodies of their human populations, with dramatic consequences for public health. As inherited sewer systems literally clog with discarded fat, the authors outline the range of policy responses being mobilized in the United States to address the problem.

In chapter 7 S. Harris Ali and Roger Keil provide a case study of an iconic global security event based on the realization that a pivotal circuit of "normal" transnational mobility was infiltrated by malign pathogens. Their analysis of the 2003 SARS crisis emphasizes the ways in which global airport and mobility flows were reorganized during the crisis in attempts to maintain global connectivity whilst attempting to interrupt the flows of the SARS pathogen. They locate the politics of the SARS crisis within the rapidly changing bordering practices of nation states. And they carefully disentangle the geographies through which global airline and other mobility systems were securitized through the use of physical architectures overlain with surveillance and tracking systems to demarcate "safe" and "risky" spaces.

The book's penultimate chapter, by the Editor, concentrates on the efforts of state militaries and nonstate terrorists, insurgents, and fighters to deliberately disrupt infrastructure networks or appropriate them as means to distribute their political violence through urban societies. Moving beyond the more usual discussions of the ways in which terrorists and insurgents target or appropriate urban infrastructure, Graham exposes how state military doctrine increasingly centers on legitimizing the wholesale destruction of urban civilian infrastructures as ways of bringing "strategic paralysis" to urbanized adversaries. Ranging from U.S. bombing theory, the systematic deelectrification of Iraq in 1991, continuing Israeli demodernizations of urban life in the Occupied Territories, and the emerging doctrine of state cyberwarfare, Graham ultimately emphasizes the importance of infrastructural warfare within the changing relations between war, political violence, and democracy.

The book closes with a chapter by Colin McFarlane which explores the links between infrastructure, disruption, and urban inequality in the Global South. Arguing that infrastructural services in many Global South cities have rarely become reliable or ubiquitous enough to be blackboxed or taken for granted, McFarlane exposes the complex infrastructural crises attendant on the remarkable recent growth of India's commercial capital, Mumbai (Bombay). He finds that continuous crises of interruption are leading to complex processes of improvisation and privatization which are tending to intensify social and geographical gaps between middle and upper classes within increasingly fortified enclaves and poor majorities inhabiting surrounding informal settlements. McFarlane's discussion of the major 2006 monsoon floods in the City, however, also demonstrate how the crisis has been exploited by elite and property interests as a reason to justify the widespread erasure of "slums" as a means of asserting Mumbai's status as a true "global" city with reliable "world class" infrastructure.

CHAPTER 2

Managing the Risk of Cascading Failure in Complex Urban Infrastructures

Richard G. Little

INTRODUCTION

Urban infrastructures are vital networks, absolutely necessary for the functioning of the twenty-first century urban complex. Modern societies and their underlying economies rely on the ability to move goods, people, and information quickly, safely, and reliably. Consequently, it is of the utmost importance to government, business, and the public at large to understand the nature of urban infrastructure and take the measures necessary to ensure that the flow of services provided by it continues unimpeded in the face of a broad range of natural and manmade hazards.

This linkage between systems and services is critical to understanding the complex relationships that exist between the physical systems and the people and enterprises they serve. Although it may be the hardware (i.e., the highways, pipes, transmission lines, communication satellites, and network servers) that initially compels attention because of its cost and spatial ambit, the public only values the services that these systems provide. Therefore, high among the concerns of those charged with managing and maintaining these systems is ensuring the continuity (or at least the rapid restoration) of service.[139]

CAUSES AND CONSEQUENCES OF INFRASTRUCTURE FAILURE

The built environment is subject to a formidable array of natural and man-made hazards. In the natural realm, earthquakes, extreme winds, floods, snow and ice, volcanic activity, landslides, tsunamis, and wildfires all pose some degree of risk to infrastructure systems. To this list of natural hazards, we can add terrorist acts, design faults, aging materials, inadequate maintenance, and excessively prolonged service lives. Although forensic analysis of past failures, some improvement in our ability to forecast wear and predict performance, and performance-based approaches to design and construction have improved the ability of infrastructure systems to withstand the effects of these hazards, failures can be expected to continue to occur.

The consequences of urban infrastructure failure can range from the merely annoying to the decidedly catastrophic. For example, whereas a water main break may only lead to a local street closure, the formation of a major sinkhole from the leakage could cause simultaneous failures in the water and natural gas systems. Fires that might result could not be fought effectively due to low flow or pressure and the property damage and possible

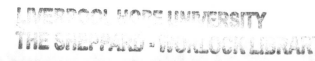

loss of life would far exceed expectations from a "simple" water main break. Historical examples of hazard events that had consequences far beyond the initial damage are the fires that followed the earthquakes in San Francisco in 1906 and Kobe, Japan in 1995 and more recently, the flooding of New Orleans following Hurricane Katrina in 2005 (Sims, this volume).

The design paradigm for protecting lifeline infrastructures has generally focused on first order effects. That is, designing the physical systems to resist the loads imparted by gravity, deterioration (corrosion), extreme natural events such as earthquakes, and more recently, malevolent acts such as sabotage and terrorism. However, as these systems become increasingly complex and modern life more dependent on them, we must also be concerned with the secondary and tertiary effects

should vital services be lost. For example, how can water services or health care delivery be maintained at acceptable levels during widespread and prolonged electrical outages?

Limiting the damage to infrastructure and ensuring continuity of service is challenging enough but further complicated by the interdependent nature of these systems. For example, almost all infrastructure systems depend on electricity. The delivery of electric power is dependent on a variety of other systems ranging from the railroads for coal deliveries, to cellular and digital communications for system control, to the public transit that workers take to the generating plant. A failure in any of these subordinate systems can cause disruptions in the electrical system with spillover effects on the others. Because of this, interdependency itself becomes a potential cause of failure. Figure 2.1 illustrates

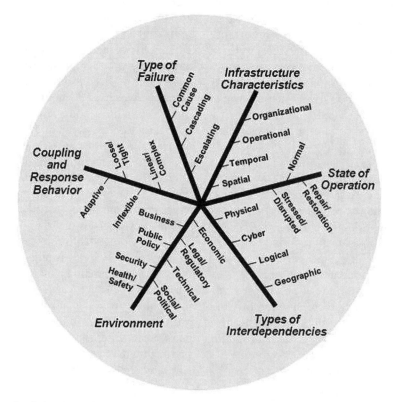

Figure 2.1 Dimensions for Describing Infrastructure Interdependencies (Source: Rinaldi, Peerenboom, and Kelly, 2001).

the complexity of interdependent relationships that can exist between various urban infrastructure systems.

Several types of interdependent infrastructure failure have been described[140] including:

- Cascading failure: A disruption in one infrastructure causes a disruption in a second infrastructure which can then disrupt a third, and so on.
- Escalating failure: A disruption in one infrastructure exacerbates an independent disruption of a second infrastructure (e.g., the time for recovery or restoration of an infrastructure increases because another infrastructure is not available)
- Common cause failure: A disruption of two or more infrastructures at the same time because of a common cause (e.g., natural disaster, right-of-way corridor)

For practical purposes, these distinctions are somewhat esoteric and in this chapter, all such failures will broadly be considered "cascading."

A cascading failure in an engineered system occurs when a failure in one of the collection of interconnected parts that delivers a service triggers the failure of successive parts. When this is translated to the case of infrastructure, cascading failure can be conceptualized as a situation where an infrastructure disruption spreads beyond itself to cause appreciable impact on other infrastructures, which in turn cause more deleterious effects on still other systems.

How far these effects propagate, and how serious they become, depends on how tightly coupled the infrastructure components are, how potent the effects are, and whether or not countermeasures such as redundant capacity are in place. Either the outage effects will die out as they move further away from the base outage, limiting overall damage, or they will gather force in successively stronger waves of cascading effects until part of or the entire infrastructure network breaks down. In the latter case, losing a key component creates a much broader failure that is out of proportion to the original failure. Given the many linkages between various infrastructure systems, cascading failures can and do cross infrastructure boundaries. Figure 2.2 illustrates how the effects of the California energy crisis propagated through other systems and the economy.

The interdependency problem is compounded by the coupling of physical infrastructure with information technology systems for communications and control. Communication and information technologies are integral components of infrastructure system design, construction, maintenance, operations, and control and more pervasive coupling of these technologies is inevitable. Applications already include coupled sensing, monitoring, and management systems, distributed and remote wireless control devices, Web-based data systems, and multimedia information systems. Although on one level, the coupling of physical infrastructure with information technology offers improved reliability and efficiency at reduced cost, the vulnerability of these systems to unauthorized intrusions from recreational hackers, hostile governments, or nonstate terrorists has led some system operators to require physical separation or "air gaps" between the system and the Internet. At the same time, experience has shown that software is fragile by nature (especially when compared to the infrastructures it controls) and there is surprisingly little known about the behavior of these coupled systems under duress. Thus, their potential to cause failures with serious consequences must be considered high.

Since the 9/11 attacks in the United States in 2001, the subsequent "War on Terror," and several high profile bombings in Madrid, London, and elsewhere, the vulnerability of civil infrastructures to terrorist attack has received considerable attention. Despite the

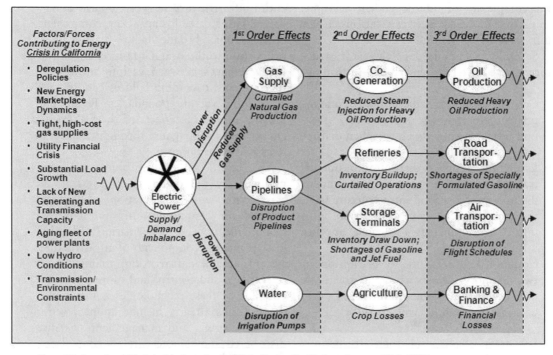

Figure 2.2 Examples of Nth-Order Interdependencies Effects (Source: Rinaldi, Peerenboom, and Kelly, 2001).

overly dramatic effects depicted in films such as the 2007 movie *Live Free or Die Hard*, the potential for failures in one infrastructure system to cause disruptions in others that could ultimately cascade to still other systems with unanticipated consequences is very real. Although some progress has been made,[141] despite considerable time, effort, and money expended to improve understanding of these systems, the linkages between infrastructures, their interdependencies, and possible failure mechanisms are still not fully understood.

There are several reasons for this that will be explored in this chapter. First, even the simplest and most straightforward infrastructure networks are relatively complicated technical systems. Computer models of these systems are approximations at best as is our understanding of their response to various insults. Further complicating this situation by coupling together the water supply system with the electrical grid and

a SCADA[142] system extends beyond reasonable limits our ability to understand precisely what happens during a malfunction. Further connecting this element of infrastructure to all the other systems on which it depends essentially defies comprehension of the resultant "system." At some point along the interdependency path, the response of these coupled systems becomes nonlinear; that is, the response to the insult is far out of proportion to the event itself. This nonlinear feedback is a key element in cascading failure. Finally, civil infrastructures are Large Technical Systems (LTS) which by their nature are intricate constructions of technology, people, and governance structures. Even if it were possible to fully model the technical components of the system, the human and organizational aspects of LTS make predicting the behavior of the system in duress with a significant degree of accuracy and precision somewhat illusory.

CLOSELY COUPLED COMPLEX SYSTEMS

In his book, *Normal Accidents,*[143] Charles Perrow described numerous failures in what he describes as tightly coupled, complex systems.[144] In the search for speed, throughput, efficiency, and the ability to operate in hostile environments, he posits that we have neglected the kind of system designs that provide reliability and security.[145] A particularly troubling characteristic of these tightly coupled, complex systems is that they predictably fail but in unpredictable ways. Because of the complex nature of the relationships that exist between system components, similar chains of events do not always produce the same phenomena. However, system level or "normal" accidents of major consequence continually recur.

Large dynamic systems of the type represented by interdependent infrastructures have been shown to self-organize into a highly interactive critical state where even minor perturbations can lead to nonlinear responses, or "avalanches" of all sizes.[146]

This work is particularly valuable to an understanding of interdependent infrastructures and extreme events because it demonstrates that certain types of failure behave in accordance with power laws rather than other probability distributions. Figure 2.3 shows the potential for underprediction of large losses in the power grid if an exponential (Weibull) distribution is assumed rather than a power law. Power law relationships imply that "…large catastrophic events occur as a consequence of the same dynamics that produce small, ordinary events." As a result, the catastrophic system failures that Perrow calls normal accidents cannot be dismissed as statistical anomalies—unique intersections of random events too unlikely to be concerned about—but rather as the expected behavior of closely coupled, complex systems. Taken together, the work of Perrow and Bak and Paczuski supports a discomforting premise that although it may not be possible to predict the precise nature of a Chernobyl, Bhopal, or New Orleans, failures of a similar magnitude and consequence are destined to occur, and probably at greater frequency

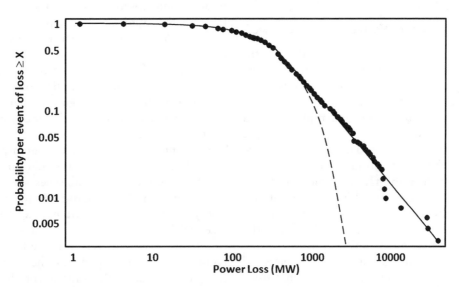

Figure 2.3 Cumulative Probability of Transmission- and Generation-Related Failures (The dashed line is an exponential (Weibull) distribution fit to the failures below 800 MW loss. The solid line is a power law fit to the data over 500 MW loss.) (Source: Talukdar, et al. 2003).

than we would like to acknowledge. Risk management under these conditions needs to be extremely aware and sensitive to the levels of uncertainty concerning failure modes and the probabilities of certain events occurring and recurring in a given sequence so that the potential for a high consequence but low probability event is not overlooked.[147]

ORGANIZATIONAL ASPECTS OF FAILURES IN LARGE TECHNICAL SYSTEMS

Although improvements to technology and a better understanding of system interdependencies are necessary to ensure the reliable provision of infrastructure services, organizations, and their internal cultures play a key role in the provision of these services. The following discussion presents the case that human capital and institutional resilience (the sociotechnological interface) are as important to the overall system as the physical assets and must be incorporated in strategies to ensure system performance.[148]

The August 2003 Northeast Power Outage

On August 14, 2003 much of the Northeastern United States suffered a massive cascading failure in the electric grid (Luke, this volume). The blackout began within the First Energy system in Ohio and rapidly spread in all directions as sections of the grid shut down to isolate the damage. There were widespread failures in interdependent infrastructures (transportation, communications, water, and sewer) and ultimately tens of millions of lives were disrupted and losses in the billions of dollars were incurred. Prior to the blackout, the vulnerability of the grid to cascading type failures was well documented.[149] The summary report of the group that studied the outage expressed no surprise in its overall findings:

Although the causes discussed below produced the failures and events of August 14, they did not leap into being that day. Instead, as the following chapters explain, they reflect long-standing institutional failures and weaknesses that need to be understood and corrected in order to maintain reliability.[150]

Regarding the cause of the initiating failure in Ohio, the Task Force continued this line of reasoning:

…deficiencies in corporate policies, lack of adherence to industry policies, and inadequate management of reactive power and voltage caused the blackout, rather than the lack of reactive power.

To a large degree the blackout appears to have been the result of corporate policies that rightly or wrongly placed the immediate business concerns of First Energy above the security of the national grid. The company never considered voluntarily cutting off power to its customers to ease congestion on the power lines and questioned why it should have interrupted power to its own customers to accommodate long-distance transmission through the system.[151] If the primary concern of a local subcomponent of the grid is to avoid shedding load unnecessarily, then more outages will inevitably occur. This may be desirable from the standpoint of the local supplier but as the August 14 blackout demonstrated, it was disastrous for the larger system.

The conflict between system reliability and profitability, both in electricity and the broader infrastructure service industries, has been a concern for many years. In 1999 an electric utility executive told a National Academies meeting that she felt that the greatest threat to reliable service delivery was the local deregulation and decoupling of the industry that would enable the subsequent divestiture of less profitable elements of the electric power grid.[152] Subsequently, the director of the Office of Science and Technology Policy made the following points in testimony before Congress:[153]

- The economy and national security of the United States are becoming increasingly dependent on U.S. and international infrastructures, which themselves are becoming increasingly interdependent.
- Deregulation and growth of competition in key infrastructures has eroded spare infrastructure capacity that served as a useful shock absorber.
- Mergers among infrastructure providers have led to further pressures to reduce spare capacity as management has sought to wring out excess costs.
- The issue of interdependent and cascading effects among infrastructures has received almost no attention.

Institutional Resilience in Sociotechnological Systems

The World Trade Center attacks of September 11, 2001 provide some interesting lessons for a more comprehensive understanding of the failure and recovery of interdependent infrastructures. Studies of the performance of infrastructure in the vicinity of the World Trade Center in the days and weeks following the attacks[154,155] underscore the notion that *resilience* or the ability to recover quickly is a critical feature of survivable systems. Resilience is often provided by means of *robustness* which increases failure-resistance through design or construction techniques or *redundancy* that provides duplicative capacity for service delivery. This work demonstrated that these characteristics are just as critical for institutions as for the physical systems themselves.

New York City was able to recover relatively quickly (compared to how other cities might have fared) after September 11 not only because of the inherent redundancy of its physical infrastructures (which is considerable) but because of its institutional resilience as well.[156] Many of the service providers involved in New York's recovery (e.g., Consolidated Edison, Verizon, AT&T, MTA) possessed considerable capacity in people considered international experts in their fields; state-of-the-art equipment and configuration management; as well as other physical and institutional resources necessary to effect recovery. This is not to imply that these service providers are unconcerned with efficiency and cost but it is not apparent that leaner, less robust systems with less of a focus on core mission and values *as service providers* would have performed as well. The evidence suggests that had these organizations been driven solely by a "lean and mean" philosophy, it is likely that recovery would have been hampered.

Normal Accidents and High Reliability Organizations

Perrow[157] was among the first researchers to discard the traditional approach to failure analysis that focused on the *technical cause* of an accident or event and the underlying *human error* that gave it life. As was discussed earlier, the technical and human elements of large technical systems are inseparable and sufficiently complex to allow unexpected interactions of failures to occur and so tightly coupled as to result in a cascade of increasing magnitude. Perrow ascribes many of the causes of normal accidents to organizational issues such as the nature of the power hierarchy and the culture of the organization itself.

Scott Sagan, in an analysis of U.S. nuclear weapons safety during the Cold War, found that although the Strategic Air Command (SAC) was essentially a tightly coupled complex system that should adhere to Perrow's normal accident theory (NAT), no accidental or unauthorized detonation of a nuclear weapon ever occurred.[158] This is ultimately attributed more to serendipity than design but Sagan emphasizes the importance (for better or worse) that organizational culture plays in the management of high risk technologies. He contrasts NAT with high reliability theory (HRT) which holds that the culture of some organizations allows them to carry out hazardous operations with a far lower failure rate than would otherwise be expected (e.g., aircraft carrier flight operations, the nuclear power industry, and air traffic control).[159,160]

Although there has been spirited debate on whether NAT and HRT are complementary or contrasting theories,[161,162,163] they both underscore the important role that organizations, and particularly organizational culture play in the management of LTS.

What is most striking about those organizations that perform surprisingly well in complex, high risk environments is an unrelenting focus on core values and an organizational culture that nurtures and supports them. By minimizing institutional conflicts, high reliability organizations (HROs) are able to strike a balance that minimizes serious technical failures while at the same time maintains reliable operations at acceptable levels. These organizations are by no means perfect and the HRO model is not applicable to all organizations, all of the time.[164] However, in all the cases described in this chapter, failure can traced to a large degree to organizational cultures that placed other objectives above the core values (i.e., safety or reliability) of the organization and which failed to comprehend fully the potential consequences of these actions. There is nothing inherently wrong with operating the space Shuttle program as efficiently as possible or in converting formerly public or publicly regulated service providers to profit-driven enterprise. However, when the pursuit of efficiency or profit becomes the motivating value of an organization whose prime responsibility should be safety or reliability, the results are likely to be unacceptable on both counts. Trade-offs between competing objectives are an organizational reality. However, these trade-offs and their consequences must be both transparent and understood. Otherwise, devastating but predictable failures (i.e., normal accidents) will continue to occur within our vital systems and the incident board convened in the aftermath will likely come to the conclusion that failure was preventable.

IMPROVING UNDERSTANDING AND LEARNING FROM FAILURE

Some form of structural failure analysis has probably existed since the time of Hammurabi if not before. Contract disputes over shoddy work or construction failures required that someone conduct an investigation and determine, as best they were able, the cause of failure and who was at fault. Forensic engineering is now a healthy, mature discipline and much knowledge has been gained, and advances made, from the study of engineering failures.[165] Engineering approaches to hazard-resistant design for structures and lifeline systems have improved continuously from the observation of past failures, assessment of their causes, and improvements in techniques and materials.[166] However, despite the value of forensic engineering to the advancement of engineering practice, the system is far from ideal. Much work of value exists only in court records, sealed by litigation settlements. Nothing analogous to the Air Safety Reporting System[167] exists for engineering practice although the Near-Miss Project at the Wharton School of the University of Pennsylvania made an attempt to develop a similar reporting framework for other industries.[168] There are also conceptual concerns with commonly used forensic techniques. In its study of errors in the health care industry, *To Err Is Human*, the Institute of Medicine noted that:

The complex coincidences that cause systems to fail could rarely have been foreseen by the people involved. As a result, they are reviewed only in hindsight; however, knowing the outcome of an event influences how we assess past events. *Hindsight bias* means that things that were not seen or understood at the time of the accident seem obvious in retrospect. Hindsight bias also misleads a reviewer into simplifying the causes of an accident, highlighting a single element as the cause and overlooking its multiple contributing factors. Given that the information about an accident is spread over many participants, none of whom may have complete information, hindsight bias makes it easy to arrive at a simple solution or to blame an individual, but difficult to determine what really went wrong.[169]

Kletz,[170] in a study of industrial accidents, also cautions about too much emphasis on causes:

If we talk about causes, we may be tempted to list those we can do nothing about. For example, a source of ignition is often said to be the cause of a fire. But when flammable vapor and air are mixed in the flammable range, experience shows that a source of ignition is liable to turn up, even though we have done everything possible to remove known sources of ignition. The only really effective way of preventing an ignition is to prevent leaks of flammable vapor. Instead of asking, "What is the cause of this fire?" we should ask "What is the most effective way of preventing another similar fire?" We may then think of ways of preventing leaks.[171]

In light of this, care needs to be taken to avoid what Taleb[172] has termed the "narrative fallacy." That is, the need for people to create a story that weaves elements of otherwise incomprehensible events into something that can be readily understood. Unfortunately, such stories usually bear little resemblance to reality and create little opportunity for learning. Lessons learned programs (or other forms of adaptive learning for understanding the causes, modes, and likelihood of interdependency failures in infrastructure systems) need to be designed to identify the influence of all contributing factors, not merely the obvious or easy ones.

The 2005 failure of the levee system that protected New Orleans from flooding provides a good case in point. Figure 2.4 depicts the sequence of events that led up to the multiple infrastructure failures that brought civil life in New Orleans to a halt. Based on the findings of several post-vent reports,[173] the failure of the levees and the resultant flooding was predictable given the nature of Hurricane Katrina. However, the event chain in Figure 2.4 indicates how the narrative fallacy could direct the focus to the levee breach or the hurricane which were just precipitating events leading up to the flooding. What it fails to capture is the root cause, the complex series of sociotechnical interactions that are embedded in the technical *and institutional* arrangements that contributed directly to the failure; that is, incorrect design basis, lack of funding, and poor maintenance.

In the aftermath of Hurricane Katrina, there

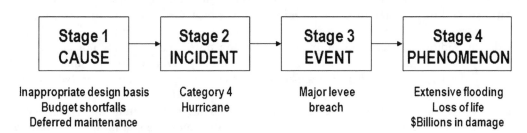

Figure 2.4 The event chain leading up to the collapse of New Orleans during Hurricane Katrina.

has been much public demand to "get the engineering right" as though that was the only problem. Although the connection between poor design, construction, and maintenance and the breach and collapse of the levees is valid, it gets caught in Kletz's "obvious cause" trap and thereby misses the point. Instead of asking: "How can we design levees so that they will not collapse or breach if subjected to a storm of Katrina's intensity?" perhaps the more appropriate question is: "How can we protect the people of New Orleans from floods in the future at reasonable cost?"

The answer to the second question lies at least as much with institutions, governance, and finance as with structural design and levee maintenance. Following Kletz's admonition, perhaps the real question is not "What are the best technologies to hold back floodwaters?" but rather, "How can we reduce exposure, damage, and casualties in the event of future hurricanes?"—which is a fundamentally different question.

These causes were abundantly apparent in the forty years leading up to the flooding of New Orleans during Hurricane Katrina (actually, the roots of this disaster are much older, predating the Civil War).[174] Beginning with the Congressional authorization of the Lake Pontchartrain and Vicinity Hurricane Protection Plan (LPVHPP) in 1965, even a cursory analysis shows a project destined to fail at some point. Federal appropriations were never sufficient to complete the project in a timely manner. As a result, construction lagged behind schedule causing further cost escalation and funding shortfalls. Local cost sharing was slow to materialize and even in-kind contributions for maintenance were not made. In addition, encroachment by local property owners made remedial work identified by the U.S. Army Corps of Engineers (USACE) difficult or impossible to undertake. Despite these obvious omissions and shortfalls, everyone involved, from the U.S. Congress to the individual homeowner, pre-

tended that that a fail-safe flood protection system was in place. Although certainly not unique, reliance on such "fantasy plans"[175] had particularly devastating consequences in New Orleans.

Assessing and Managing the Risk of Cascading Infrastructure Failure

Risk is a useful analytical concept that gives meaning to those uncertainties of life that pose a danger to people or what we value.[176] Risk is often expressed as a combination of the likelihood of an adverse event, the vulnerability of people, places, and things to that event, and the consequences should that event occur; that is, the probability of an adverse event (threat and vulnerability) multiplied by the consequences of that event, or $R = P \times C$. For example, if we consider the case of rising sea level, the risk is greater to people living in coastal areas than to those at higher elevations because of their increased vulnerability to lowland flooding and the greater consequences (to them) if flooding occurs.

One of the inherent shortcomings of this simplified, expected value-type approach to risk is that the structure of the model can produce apparently similar risk levels for vastly different classes of events. For example, from an arithmetic standpoint, a catastrophic event with extremely low probability can be interpreted to have a similar level of "risk" as a relatively frequent event with far lower consequences. Although it is compelling to believe that designing for some "maximum probable event" (or Standard Project Hurricane in the case of New Orleans) fully addresses the risk issue, the experience with Hurricane Katrina demonstrated that this may not be the case. In addition, relying on the presumption that protection from a "maximum probable event" has been provided when this is not actually the case will increase the vulnerability (and hence the risk) if more people locate in a formerly hazardous area now believed to be safe.

A more formalized process of risk assessment and risk management can help to illuminate and deal with these uncertainties at least from the standpoint of understanding the value of potential mitigating actions. *Risk assessment* has classically been defined by three questions:[177]

1. What can go wrong?
2. What is the likelihood that it could go wrong?
3. What are the consequences of failure?

As has been discussed previously, beyond possible initiating events (e.g., earthquake, hurricane, terrorist attack) and first order effects, it is extremely difficult if not impossible to describe accurately the failure modes that could come into play during the cascading failure of interdependent infrastructures.

At the same time, the tendency for LTS to exhibit the self-organized criticality and nonlinear failure cascades demonstrated by Bak and Paczuski does bring into question the likelihood of developing realistic probability estimates for events that occur at the tails of probability functions. These are the high-consequence "black swans" that Taleb describes in his eponymous work on highly improbable events.[178] For the cascading failures that are the subject of this chapter, it appears that a focus on consequences could yield more useful management guidance.

Based on the findings of the post-Katrina assessments,[179] it is clear that in the case of New Orleans the answers to these questions could have been summarized as follows:

In the event of a not uncommon intensity hurricane, it is highly likely that the levees will be breached or otherwise fail in a number of locations with the result that hundreds to thousands of mostly poor people will perish and damage in the billions of dollars will accrue.

Had this perfunctory statement of risk been broadly disseminated, it is interesting to speculate on whether it would have affected development patterns in the city, especially in particularly vulnerable areas such as the Ninth Ward. However, not even this pessimistic forecast needed to preclude all activity in vulnerable areas. Risk can be managed, and, under the appropriate conditions, managed quite effectively. Risk management is the process by which the results of risk assessment are integrated with other information—such as political, social, economic, and engineering considerations—to arrive at decisions about the need and methods for risk reduction. Risk management seeks answers to a second set of questions:[180]

4. What can be done and what options are available?
5. What are the associated trade-offs in terms of all costs, benefits, and risks?
6. What are the impacts of current management decisions on future options?

It is the answers to these three questions in the context of governance and decision making in New Orleans that will attempt to provide some insight into what can be done to manage cascading failure of urban infrastructure.

What Can Be Done and What Options Are Available?

Options for managing flood risk can be grouped into five general categories. We can

1. Avoid the risk by locating somewhere else. In the case of flood protection, not locating in the flood prone area is perhaps the wisest choice. However, for many communities this is no longer an option.
2. Reduce the risk by taking countermeasures. These might include a combination of advance warning and evacuation, flood protection works, such as levees, hazard-resistant structures that are designed with sufficient robustness or sited

above likely flood levels, and rapid response and recovery mechanisms so that the city has sufficient resilience to recover quickly.

3. Spread the risk by choosing multiple redundant locations for certain activities. (It would have been wise to locate at least some of the special generating capacity for New Orleans' dewatering pumps outside of the flood zone.)[181]

4. Transfer the risk by buying insurance. The relevance of this option will depend on the continued viability of the Federal Flood Insurance Program and the willingness and ability of the commercial insurance industry to underwrite flood hazard risk at rates that homeowners are able and willing to pay.

5. Retain the risk. In light of the preceding points, property owners, neighborhoods, the City of New Orleans, or the entire state of Louisiana may have no choice but to accept a portion of the consequences of the multiple hazards they face. Catastrophe bonds,[182] either private (insurance industry) or sovereign (local, state, federal) may be an option to supplement traditional insurance. Mexico has recently issued a $160 million catastrophe bond to cover losses in the event of an earthquake greater than a prespecified magnitude and the declaration of a national emergency.[183]

What Are the Associated Trade-Offs in Terms of All Costs, Benefits, and Risks?

Benefit/cost (B/C) analysis was practically invented to do the type of trade-off analysis inherent in large flood control projects. A very simplified B/C analysis would determine that in exchange for $X in capital outlays and $Y for annual operating and maintenance expenses, benefits totaling $Z would be estimated to accrue. If the net present value of the annualized equivalent of $Z > X + Y$, the project has a favorable cost–benefit structure

and is justified. However, this analytical procedure makes no effort to distinguish between who bears the costs and who reaps the benefits. For example, although all U.S. taxpayers underwrite a portion of the federal share of the costs of flood control, the benefits accrue locally. Although often labeled "National Economic Development" benefits, these are usually targeted to reach a far narrower audience. The equity of federal water development has been debated for years and will not be resolved here. However, it will suffice to say that from the standpoint of managing risk equitably, much better alignment of who benefits and who pays is certainly possible.

WHAT ARE THE IMPACTS OF CURRENT FLOOD MANAGEMENT PRACTICES ON FUTURE DEVELOPMENT OPTIONS?

The Mississippi River valley historically has been a focal point of human's efforts to "harness" nature to their purposes. The U.S. Army Corps of Engineers and its predecessors were charged with controlling the periodic and damaging floods that occurred on the Mississippi River and its tributaries. New Orleans, subject to both river flooding and tidal storm surge, has also seen the almost continuous installation of flood works. The decision to encourage the growth of a major city and build flood works to enable that growth precluded other approaches. Once the size of the population and the value of the constructed environment achieved certain thresholds, there was little to do but to keep investing in large protective flood control structures. However, as demonstrated by Hurricane Katrina, these large investments were not sufficient to avoid disaster.

At this point, New Orleans and the United States are faced with another set of decisions that will affect the city and its residents far into the future—to continue with the failed policies of the past or to seek a more harmonious and equitable balance with the forces of nature, the desires of man, and basic economics.

Hurricane Katrina demonstrated what can happen when the risk management process is manipulated to produce a comforting but inaccurate depiction of likely events. Actions taken to develop and implement comprehensive hazard mitigation strategies for infrastructure must be based on a balanced assessment of all risks confronting the systems and the possible consequences of their failure, either singly or in combination with other, interconnected systems. These strategies must be informed by the best available information and carried out by people knowledgeable about the systems, their possible failure modes, the implications of concurrent system failures, and possible interventions that would allow systems to degrade gracefully and avoid catastrophic, multisystem failure.

CONCLUSIONS

Urban infrastructure faces a range of potentially serious threats. In addition to natural hazards, experience demonstrates that excessively prolonged service lives, aging materials, and inadequate maintenance all negatively affect infrastructure. Despite this formidable array of threats confronting our infrastructures, many problems will occur simply due to the complexity of these systems. Potential failure nodes are repeatedly created at the intersections of tightly coupled, highly sophisticated transportation, electric power, and telecommunications systems and are compounded by their reliance on information systems and software. As a first step in protecting these systems, the "vulnerability of complexity" must be resolved. Although, there is strong capability within the infrastructure and hazard communities for identifying and assessing these vulnerabilities, without a better understanding of the overall context in which they need to be applied, vulnerability assessment represents only part of a total systems solution.

Our basic systems are at risk from threats we may not yet foresee. Prudent risk management demands that we anticipate these threats to our physical infrastructures, design systems that are inherently safer and more resilient, and be prepared to restore them when they fail.

CHAPTER 3

Disoriented City:
Infrastructure, Social Order, and the Police Response to Hurricane Katrina[184]

Benjamin Sims

From a technological point of view, what is unique about cities is that they are dense concentrations of infrastructure. In the core of most cities, the infrastructure of roads, buildings, and parking lots visually dominates the landscape, to the near complete exclusion of the natural environment. Housing and transporting a large urban population involves the construction of massive buildings, bridges, tunnels, and other civil engineering works. A dense tangle of infrastructure is embedded within these massive structures, extending deep underground as well as high above the streets.

Cities are also complex, highly structured social institutions. Much of the institutional structure of cities is built around infrastructure networks. Distinct bureaucracies may oversee roads, electrical power transmission, sewer systems, building construction, and the protection of buildings from fire. In addition, elements of city government that aren't directly associated with infrastructure are highly dependent on the city's infrastructure simply because they are located within the geographic boundaries of the city. Finally, the employees of all of these organizations live or work in the city, and like all city dwellers, depend on infrastructure to support their activities and to orient themselves in the space and time of the urban environment.

New Orleans is no exception to this urban dependence on infrastructure. The urban core of New Orleans is a densely populated, older urban area crisscrossed by multiple interdependent infrastructure networks, many in a state of decay. In addition, most of the city lies below sea level and continues to sink, meaning that its continued existence as a populated place is entirely dependent on an infrastructure of levees, canals, and pumps that keep water out of its bowl-like terrain.[185]

Hurricane Katrina had an astonishingly swift and devastating impact on infrastructure along the U.S. Gulf Coast, but its most concentrated impact on infrastructure and human life was in New Orleans.[186] The hurricane roared ashore beginning the evening of August 28, 2005, its eye wall passing near New Orleans on the morning of August 29. As the eye wall came ashore, the electrical system, the telephone system—both wired and wireless—and Internet connections were destroyed or severely damaged. Some towers used for emergency response radio systems were also knocked out by the storm.[187] Also that morning, levees and floodwalls surrounding New Orleans began to fail, and the city started to slowly fill with water (Figure 3.1).[188] By the next morning, the city was inundated, and with that inundation, most roads became impassable and sewers no longer functioned;

Figure 3.1 New Orleans infrastructure, underwater: Interstate 10 at West End Boulevard, view toward Lake Pontchartrain, August 29, 2005, shortly after the passage of Hurricane Katrina. U.S. Coast Guard photograph by Petty Officer 2nd Class Kyle Niemi. Source: U.S. Coast Guard.

at some point, running water was shut off.[189] Homes, hospitals, and police precinct buildings were flooded. In a final blow, as the water rose, it swamped the backup generators powering many of the remaining radio towers used by local agencies, causing them to shut down.[190] At this point, local emergency responders were essentially without remote communications of any kind, and had severely limited mobility and supplies.

Hurricane Katrina also destroyed elements of infrastructure specifically connected with the New Orleans Police Department (NOPD)(Figure 3.2). The department lost access to many of its facilities in the flooding, including its headquarters, crime lab, evidence room, armory, jail, and many police stations.[191] This left the police short of supplies, including food and ammunition. About a quarter of the department's police cars were also lost or stranded in the flooding that resulted from the failure of the levee system;[192] in any case, many areas of the city were impassable to standard vehicles due to debris and flooding, and it became impossible to obtain gasoline.[193] Like other local agencies, the police department also lost its main communications infrastructure when radio antennas were damaged in the storm. Police

radios could still be used in walkie-talkie mode to communicate with nearby units, although the frequency they operated on was soon overwhelmed with traffic, making communications difficult. In any case, the radios' rechargeable batteries were soon drained and there was no electrical power available to charge them.[194] Alternative communications technologies, such as personal cell phones, were not working either.[195] Finally, as if these problems were not enough, New Orleans police officers were required by law to live within the city. The storm left almost 900 of

Figure 3.2 Police cars and emergency vehicles, presumably parked on an overpass to keep them out of floodwater, ended up stranded: New Orleans, September 7, 2005, a week after Hurricane Katrina's initial impact. U.S. Federal Emergency Management Agency photograph by Jocelyn Augustino. Source: U.S. Federal Emergency Management Agency.

the department's approximately 1,400 officers homeless, leaving them without their normal off-duty sources of food, shelter, and clothing. Many officers, even those who did not lose homes, were also preoccupied with their families who had evacuated the city.[196]

The loss of virtually all of the police department's own infrastructure, with no effective contingency plans in place, thoroughly compromised the department's normal organizational routines. Officers lacked the mobility and communications capacity to patrol or effectively coordinate response to crimes. With homes and police stations unusable or destroyed, they lacked supplies, even for basic needs like food, shelter, and bathing.

In the aftermath of the hurricane, there was a strong sense within the city that social order was collapsing. Rumors of social breakdown were rampant, and were spread via uncritical media accounts. In particular, the Superdome—where many residents of the city who did not leave town were housed during and after the hurricane—became the object of rumors of widespread intimidation, rape, and murder. Even high-level officials were not immune from spreading these rumors: Police Superintendent Eddie Compass appeared on *The Oprah Winfrey Show* shortly after the hurricane, tearfully claiming that "'little babies' were being raped in the Superdome."[197] Though these rumors of extreme acts turned out to be greatly exaggerated, they functioned as a kind of shorthand for the overwhelming sense of social collapse felt by those remaining in the city, a sense that all the normal institutions of city life had simply ceased to function.

The sense of social order coming undone was as strong within the NOPD as it was anywhere in the city. Police buildings and police cars were destroyed, criminals traveled the city with impunity, often shooting at the police, and police officers lacked the resources to rescue people or even dispose of dead bodies. In many instances, police were very resourceful in improvising solutions to these problems, putting together makeshift command centers in dry parking lots, and in several cases coordinating informal boat rescue efforts. Nonetheless, the police department is widely reckoned to have failed as an institution in the aftermath of the storm, becoming ineffective at controlling crime or managing rescue efforts. Hundreds of police officers resigned or simply walked off the job, two committed suicide, and those who remained were severely demoralized and under extreme stress.

Because the NOPD is a well-defined, clearly bounded social institution that had a particular role to play in the response to the storm, and because its problems were widely reported, it is an ideal microcosm for exploring the collapse of urban social institutions and social order in the aftermath of Hurricane Katrina. As an emergency response organization, it is also one we would hope would be able to function following a major disaster. Its problems following the hurricane raise some critical issues about how the United States, as a society, has sought to prepare for natural disasters and attacks—the objects of what we now call Homeland Security.

INFRASTRUCTURE AND SOCIAL ORDER

The question of social order—that is, how human beings are collectively able to enact regular social structures and patterns of behavior—is central to the social sciences. While there are many ways of describing social order, the perspective I use here derives from the classic Durkheimian tradition in sociology and anthropology, in particular as it has been developed in the work of anthropologist Mary Douglas.[198] It also draws from the science and technology studies literature. This literature has particularly emphasized the idea that social order is embodied in material and technological artifacts;[199] infrastructure is seen as unique in the way it is connected to social order across large spans of time and space.[200] Drawing from these sources, social order can be defined in terms of a number of interrelated elements:

- *Categories:* these include basic definitions and distinctions among social and material objects; for example, classes of things, like animal and plant, or social identities, including job categories like police officer.
- *Rules or norms* pertaining to those social categories: Certain norms, for example, might define how someone occupying the position of police officer should behave. A particularly powerful set of rules are pollution rules, which define certain actions, particularly those that involve the disruption of boundaries between categories, as dirty or dangerous.
- *Practices* that draw on and reinforce those categories and rules: For example, there are numerous practices for removing potentially dangerous or dirty elements from places they don't belong, such as bathing, snow removal, or pumping away floodwater. Social rewards or sanctions may serve to reinforce individual respect for categories and compliance with rules.
- *Durable structures*, both social and material, that draw on and reinforce the aforementioned categories, rules, and practices: Some practices become institutionalized, persisting in similar form for long periods of time. In many cases practices are made durable by being embedded in material things. Built barriers, for example, can serve as durable means for keeping both people and material objects in their socially acceptable places.

Technology plays a key role in each of these elements of social order. We use technologies to classify objects in the world through measurement and standardization. Technologies also serve to enforce social categories and constrain practices. Bruno Latour, for example, has shown how objects as simple as keys, doors, and child restraints can engage their users in a complex ballet of delegations of competencies between human and technological entities.[201] This process is normative in the sense that it enforces upon the user a certain limited set of roles and possible courses of action. Finally, if certain sociotechnical arrangements become stable over time, they become part of the durable social order. Infrastructure is a key part of the durable social order because it structures practice and provides common reference frames over much wider spans of time and space than other technologies. Indeed, it is difficult to explain social order on the scale of modern societies at all without reference to infrastructural technologies.[202]

BREAKDOWNS IN SOCIAL ORDER

In the aftermath of Hurricane Katrina, the NOPD experienced what organizational sociologist Karl Weick has termed a "cosmology episode," a situation in which "people suddenly and deeply feel that the universe is no longer a rational, orderly system."[203] A cosmology episode occurs when all of the key elements of social order mentioned above break down, to the point where even the fundamental categories that the impacted people use to organize the world are called into question—in other words, their cosmology. Under such circumstances, organizations may experience difficulties with sensemaking—the ability of individuals within an organization to make their actions meaningful in relation to a wider organizational view of reality.[204] When this capacity breaks down, organizations find it difficult to come up with collective representations of situations, which in turn can make it difficult for them to take collective action to respond to events.

The social science literature on disaster has noted that difficulties with sensemaking following disasters are often rapidly overcome, resulting in creative improvisation and recombination of available resources to solve problems.[205] In the immediate aftermath of Hurricane Katrina, however, the NOPD was never able to fully overcome the sensemaking difficulties it encountered, resulting in a diminished problem-solving capacity. Sensemaking did not completely break down. A number of cases of effective improvisation by the police are evidence of this. These included assembling makeshift command centers, taking control of stores to obtain food, water, and needed equipment, pumping gasoline from underground tanks using jury-rigged electric pumps powered by generators and car batteries, and organizing boat rescue efforts.[206] However, the ability of the police department to improvise appears to have been largely exhausted by efforts to obtain resources for its own survival, with little capacity left over to coordinate rescue or crime-fighting activities at an organizational level.

The ability of the NOPD to respond to Hurricane Katrina was compromised by three major breakdowns in social order: a breakdown in the barriers that separate the human body from pollutants; a breakdown in the spatiotemporal routines of police work; and a breakdown in the moral order of police work. Each of these breakdowns was the direct result of infrastructure failure, and each called into question the elements of social order described above: durable structures, practices, rules and norms, and fundamental categories. This, in turn, led to difficulties with sensemaking, including a profound sense of spatiotemporal disorientation as well as a breakdown of morale and commitment to the department.[207]

BARRIERS BETWEEN THE HUMAN BODY AND POLLUTION

When the levees around New Orleans failed, they let massive quantities of water into the streets. With the failure of this one infrastructure, it became impossible to put into practice the normal rule of separation between water and roadways. This was important because it immediately made many streets impassable to conventional cars and trucks. But the water transgressed other boundaries as well. Other substances are readily carried by water, and because water is a fluid, it readily fills spaces and crosses material barriers when given the opportunity to do so. The floodwater in New Orleans immediately dissolved the boundary between the sewage system and the rest of the city, picking up waste and toxic substances and distributing them around the streets.[208]

Mary Douglas suggests that our modern concept of dirt is essentially one of "matter out of place":

> Shoes are not dirty in themselves, but it is dirty to place them on the dining-table; food is not dirty in itself, but it is dirty to leave cooking utensils in the bedroom, or food bespattered on clothing; similarly, bathroom equipment in the drawing room; clothing lying on chairs; out-door things in-doors; upstairs things downstairs; under-clothing appearing where over-clothing should be, and so on.[209]

By this standard, the failure of the levee infrastructure, and resultant flooding, immediately made the streets of New Orleans a very unclean environment (Figure 3.3).

One of the key sources of information on the situation of the NOPD after Katrina is the reportage of Dan Baum, who spent time with a number of police units in the weeks following the storm. New Orleans police officers, like many others forced to get around in the flooded portions of the city, were unable to avoid contact with the dirty water in the streets, as illustrated by the following vignette from Baum:

> Captain Edwin Hosli...sloshed through filthy knee-deep water at the corner of Napoleon Avenue and Carondelet Street, directing his own makeshift [boat rescue] operation. The smell of rot and gasoline that blanketed the city was strongest here at the water's edge, where float-

Figure 3.3 Matter, and infrastructure, out of place: Lower 9th Ward, New Orleans, September 18, 2005, three weeks after Hurricane Katrina's initial impact. U.S. Federal Emergency Management Agency photograph by Andrea Booher Source: U.S. Federal Emergency Management Agency.

ing sewage and garbage gathered in foamy skeins.… He carried a black semi-automatic assault rifle and wore the same squishy wet shoes and uniform he'd had on since the storm.[210]

As a result of this constant exposure to wet and polluted water, many police officers developed blisters, rashes, and fungal infections on their feet and legs.[211] To make matters worse, lack of electricity and scarcity of water, along with lack of access to homes, meant that there was no easy way for officers to bathe or wash their clothes—two actions that could have served to restore the usual separation between bodies and dirt. Some improvisation occurred: one officer Baum talked to washed her uniform daily in a waste basket.[212] Another refused to wear his clean uniform shirts:

In the car where he sleeps, he has hung three starched, white uniform shirts wrapped in plastic. He's been wearing a ratty, gray T-shirt for several days.

"I'm not going to wear those starched shirts in this filth," he explained. "I'm saving them."[213]

As this instance suggests, not only bodily comfort but some aspect of the identity of police officers was threatened by exposure to the polluted environment of post-Katrina New Orleans, to the extent that this officer found more value in preserving some of the symbolic trappings of the police role in a pristine state than in wearing his shirts.

In erasing the boundaries between the sewage system and the streets, the levee failure also rendered the sewer system inoperable. This, in turn, made it impossible for those remaining in New Orleans to manage human waste according to normal standards. The problem was compounded by the fact that police officers, working out of improvised command posts, had almost no space to carry out any personal hygiene activities. One imagines there must have been a considerable amount of improvisation in this area, considering the circumstances, though this has not been documented in media accounts.

The inability to maintain normal social order around the body and things regarded as dirty or polluted was cited as a major difficulty by the police force. Police Superintendent Compass, with characteristic hyperbole, put it this way: "If I put you out on the street and made you get into gun battles all day with no place to urinate and no place to defecate, I don't think you would be too happy either."[214] By a few weeks after the hurricane, the department was able to house its officers on a cruise ship. Even with two officers to a room, this was considered, as one officer put it, a "lifesaver": "If it wasn't for that, being able to eat a hot meal, having a place to stay, I think I would have lost my mind."[215] The cruise ship presumably also provided shower and bathroom facilities.

Douglas suggests that the human body frequently serves as a metaphor for social order in general. In modern societies, one path between the integrity of the body and the integrity of social order is through infrastructure: because the rituals surrounding bodily waste and cleanliness are intimately entangled with large-scale infrastructures, destruction of those infrastructures can directly impact our sense of our bodies and their relationship to the world. In turn, this impact on bodies may serve as a powerful amplifier of individuals' sense of the scope and severity of breakdown in infrastructure and social order. This may be one reason why the inability to maintain order around the body had such an impact on morale in the NOPD.

Spatiotemporal Routines

Another effect of infrastructure destruction was to radically alter the spatial and temporal order of the city and of police work

specifically. The space of police work was disrupted at two levels. First, the infrastructure of police buildings was largely destroyed or rendered unusable, leaving the police no place to enact many of the daily routines and interactions that serve to coordinate police work and create a shared sense of identity and purpose. The importance of this function is reflected in the fact that much of the improvisation that occurred within the department following Katrina had to do with setting up command posts wherever possible—store parking lots, casino driveways, schools, nursing homes, hotels.[216] The hurricane also profoundly disordered the city landscape in which police officers were used to operating. Flooding rendered many streets impassable to vehicles, making it necessary for police to find new routes, which frequently forced them to take the wrong way down one-way streets or highways. Some officers apparently found that their knowledge of the streets was not easily "reversible" in this way, adding to the difficulty of getting around.[217] Those involved with boat rescues also had difficulty navigating streets where familiar landmarks were destroyed or submerged.[218]

The disruption of other institutions by the flooding also impacted police routines. Baum relates the story of one officer who spent the better part of a day driving around with a dead body in his patrol car, unable to find anyone willing to deal with the body. After being unable to contact the coroner's office, and being turned away from local emergency rooms, he was forced to return the body to the street where he found it.[219]

A sense of mingled spatial and social disruption is captured vividly in another description from Baum:

The boat proceeded slowly up Napoleon Avenue, bumping against sunken cars and fallen trees. Graceful multicolored turn-of-the-century houses reflected prettily in the calm water. The officers ducked a street sign as they rounded the corner onto commercial Claiborne Avenue, and fell silent as their view widened to a panorama of their city. A body floated face down in a used-car lot. The rounded shoulders of another bobbed near a funeral home. The giant root-beer mug that announced Frostop Burgers was upside down and half submerged. On the horizon rose a thick spiral of heavy smoke. A young woman sunbathed on top of a heap of boxed toasters, blenders, and other kitchenware piled into a speedboat moored outside a Walgreens drugstore. She waved nervously and yelled to someone inside the store; the cops cruised past.[220]

The temporal order of police work was also thoroughly disrupted by the hurricane. Instead of a normal routine with distinct work shifts and time spent at home, police officers were essentially on a work footing all the time, even when they had a chance to sleep. This was tied to the destruction of the space of police work and the city—with homes destroyed and mobility restricted, standard temporal as well as spatial routines became impossible.

Moral Order

The essence of police work is the enforcement of moral order: distinguishing between right and wrong actions, and making sure people who cross that boundary are stopped or removed from society at large. The practices that enable this sorting depend on infrastructural support in the form of police stations, jails, and prisons. Another aspect of the enforcement of moral order is that police officers are supposed to assist the innocent, protecting them from criminals and keeping them from harm. This also depends on the infrastructure that enables police mobility and communications. As enforcers of moral distinctions, a key part of the role of the police officer is to serve as an exemplar of unambiguously right behavior. The ability to behave righteously, however, is itself dependent on elements of infrastructure; for example, on the supply chains and monetary exchange systems that make it possible to

clearly distinguish between legal and illegal ways of obtaining property. Because it damaged or destroyed the infrastructure attached to these aspects of moral order, Hurricane Katrina threatened the moral standing of the police department.

For the police, one of the more frightening aspects of the posthurricane environment was the sense that they were being outgunned by criminals who were no longer reluctant to fight back against the police. Rather than running at the sight of police lights, looters were apt to shoot at police cars.[221] Nor was it clear what police officers were to do with offenders if they did capture them, with the normal jail and prison infrastructure out of commission or inaccessible from inside the city. One officer took two looters into custody and handcuffed them to a railing at the hotel where he was staying, but had to let them go because he had nowhere to send them.[222] Other officers simply photographed looters with the items they had stolen and let them go, hoping to be able to catch them later on. Ironically, the Louisiana Department of Public Safety and Corrections had set up an improvised jail at Union Station in New Orleans, but apparently officers on the street did not know of its existence.[223]

Some police officers were also disturbed about their inability to help victims of the storm. Given the scope of the need, the difficulties in getting around the city in police cars, the lack of medical facilities, and the relative scarcity of boats, many officers were forced to stand by ineffectually while people suffered and died. One officer related the following horrifying incident to a reporter:

> He was helping at the convention center one night when a man came up to him carrying his baby in a filthy blanket.
>
> "The baby's lips were blue," he remembered. He hadn't eaten in days, and his mother was unable to breast-feed because she was ill.
>
> [He] didn't know what to do. There was no hospital, no paramedics to call. He rushed the father and baby into his car, and began speeding

west, away from the water. He stopped in St. Charles parish and called an emergency medical service crew, which picked up the child. He found out later that the baby did not survive.

"I never thought in my wildest fears that this could happen—that a baby could starve like that in America. I have to think God has a reason," he said.

A few days later, after the National Guard arrived, [the officer] saw a huge pallet of baby formula at the police headquarters and was in agony all over again.[224]

On a similar note, one of the officers who committed suicide was said to be upset because "he couldn't help stranded women who were pleading for food and water" and could not rescue trapped animals.[225] The role of police as helpers of the innocent is often a matter of facilitating the handoff of problems to the appropriate authorities outside the police department. With the destruction of infrastructure, police lost access to many of those outside institutional resources and found that, without this support, they were often unable to help in ways they felt were adequate.

The NOPD has long been notoriously corrupt, so the ability of its officers to draw on the cultural image of police as moral exemplars has never been particularly strong. The exigencies of operating in the posthurricane environment put their moral status under additional strain. For one thing, there were a few clear-cut cases of police officers participating in looting, taking over space and scarce resources for personal gain, and threatening those who questioned their actions.[226] These were isolated cases, but they threw many of the actions of other police officers into question. Many of the actions the police had to take to obtain needed resources to continue their work were morally ambiguous. With the normal infrastructure of financial transactions essentially destroyed, the police were forced to commandeer supplies from stores within the city. Without effective communications, the police command struc-

ture was not functional, making it necessary for individual units to make their own decisions about what kind and quantity of material were appropriate to commandeer. In such an environment, it was not always easy to distinguish looting from legitimate efforts to procure supplies. These ambiguities have largely been resolved since the storm through a series of investigations in which the activities of some officers were prosecuted while others were legitimized.[227] But in the immediate aftermath of the storm, no such structures for legitimizing actions were in place, forcing officers to rely on their own moral reasoning in an inherently ambiguous situation. Given the circumstances, it certainly seems possible that even well-intentioned officers could have made decisions that, in hindsight, appear to be wrong.

IMPACT OF BREAKDOWNS ON THE NOPD

The ability to orient the self and the group in time and space is a key element of sensemaking. Without this orientation, it can become difficult to coordinate activities or measure accomplishments, making it difficult to maintain social order.

Disorientation in Space and Time

In modern societies, infrastructure is one of the key means we use to orient ourselves in space and time. Spatially, infrastructure—especially transportation infrastructure, like roads, walkways, and train lines—is a network that makes connections between some points relatively easy, and others much harder. For example, it might be easier to move between parts of a city that are along the same major roadway than it is to move from one of those places to an area only accessible by side streets, affecting our perception of distance between those places. One-way streets can similarly affect perception of urban space, as New Orleans police officers found out when they tried to navigate without that restriction. A large-scale disruption to these transportation networks can therefore render urban space difficult to navigate. Infrastructures have a more subtle spatial orienting function as well—for example, a city resident might think of direction in terms of an iconic bridge or skyscraper, or a particularly distinctive overpass or intersection. This kind of orientation is often primarily visual, so even subtle changes in visual texture— such as from an inundation of floodwater— can make a city seem unfamiliar.

The infrastructure of time keeping is fortunately portable enough, thanks to wristwatches, that it remained available in the aftermath of Katrina. But infrastructure also structures time in more subtle ways, by enabling many of the practices and rituals that punctuate daily routine and orient people in time: bathing, eating, moving between home and work. For New Orleans police officers, infrastructure destruction made these activities difficult or impossible, or shifted them in time, creating a subjective feeling of one unending day.

For example, in the midst of driving around the city with a dead body, one of the police officers mentioned above reported a feeling that time had come to a standstill:

> He parked a few blocks from the Superdome, staring through the windshield at the huge structure rising incongruously from deep water. "I was dazed and confused," he told me later. All he had was his uniform, the cash in his wallet, and his gun. He didn't know what to do with the corpse. The entire edifice of city government seemed to have dissolved in the floodwaters. He sat gazing at the Superdome for two hours.[228]

Several comments by police officers in newspaper and magazine reports corroborate this sense of temporal disorientation. One sergeant was quoted as saying: "this is the same kind of stress I experienced in war… there is no way to tell what day of the week it is…time just stopped on August 29th and it has been one long, continuous day ever since."[229] Warren Riley, Compass's successor as police chief, told a reporter "it disturbs me that I don't know where I slept for the first few days…it reached a point when we didn't know what day it was."[230] Likewise, a Louisiana State University psychologist who worked with the NOPD after the storm overheard someone in an elevator ask a police officer what day it was. The officer reportedly responded, "I know what day it is. Every day is the same day; it's the day after the hurricane."[231]

Decline of Morale and Commitment

In the days after the hurricane, signs started to emerge that the police department was in deep crisis as an organization. Two police officers committed suicide, one of them apparently despondent over being unable to help people and animals trapped by floodwaters.[232] Ninety-one officers resigned in the days and weeks after the storm, while 228 were investigated for simply leaving their posts without permission.[233] These departures suggest that, for many officers, commitment to the police department, and the role of police officer, could not compete with other commitments, such as personal survival or taking care of family. Baum quotes one officer who left as follows:

> Look, man, I stayed that whole week…no electricity, no radio communications. I hadn't heard from my wife and kids…. I finally decided this, this job…." He sighed, looking for words to describe the thanklessness of being a New Orleans cop. "I decided that my family was more important."[234]

This officer's comments move directly from infrastructure problems, to family issues, to the difficulties of police work, suggesting the close connection between infrastructure and these commitments. This kind of calculation appeared to be widespread; Police Superintendent Eddie Compass resigned shortly after the hurricane, and was quoted by Mayor Ray Nagin as saying "look man, I've done my share," citing his need to take care of his pregnant wife as a reason for resigning.[235]

The departures and suicides appear to have been connected to an overwhelming sense of failure within the police department, connected to the lack of infrastructural resources and consequent inability to perform police work. One of the officers Baum interviewed, who stayed for the duration of the crisis, described his sense of the outcome:

> "Today, it finally hit me," he said softly. "I woke up and thought, there's nothing here for me. Not at work. Not at home. What did we accomplish? Nothing. We took such an ass-whipping. We didn't stop the flooding. We didn't stop the looting. The whole city got destroyed. We lost."[236]

While the police department, as an organization, was crippled at a purely functional level by the loss of basic infrastructure, there seems little doubt that these second-order failures of sensemaking and morale, themselves a result of infrastructure problems, were what brought the NOPD to the brink of complete collapse.

CONCLUSION

The swift collapse of, essentially, the entire infrastructure of the city of New Orleans exposed an uncomfortable truth: existing infrastructure, even in a relatively wealthy and highly technological society like the United States, is surprisingly vulnerable to destruction—not only because specific systems are not robust, but also because these systems are thoroughly and often perversely interdependent. More seriously still, it showed that cities, as social institutions, can be exceedingly vulnerable to infrastructure collapse. In the immediate aftermath of Katrina, city government almost entirely ceased to function: decision makers were in danger and cut off from their supporting organizations, employees of many city organizations had been allowed to leave the city, key facilities were damaged or destroyed, and emergency response agencies, like the police department, found themselves without the resources to effectively manage the situation. These difficulties were compounded by the failure of state and federal officials to fully comprehend the devastation in New Orleans.

Cities count on the efforts of emergency response agencies to maintain social order in the face of constant low-level threats, as well as the occasional major disaster. This creates a characteristic dilemma: to effectively respond to all kinds of threats to the urban fabric, emergency responders must themselves be highly embedded in this environment and have intimate local knowledge of its workings. Having such a close working relationship with the urban environment, however, means that the routines and practices of these agencies become highly tuned to local circumstances, and as a result are more vulnerable to disruption in a disaster that affects the city.

The response of the NOPD to Hurricane Katrina reveals an organization crippled by the unavailability and destruction of infrastructure, at more than one level. At an immediate level, the capacity of police officers to communicate at a distance, to move about the city, and to obtain needed resources was severely limited. This, in itself, partly explains the problems the NOPD had in responding to the disaster. But it does not explain police officers walking off the job, committing suicide, or becoming demoralized to the point of despair, which may have had an equally severe impact on the ability of the department to respond effectively. These secondary effects were directly related to the immediate impacts: the department lost capacity to perform key policing tasks, and with this, police officers were unable to enact key elements of police identity and culture, calling their role as emergency responders and public servants into question. At the same time, the flooding made it difficult to maintain boundaries around the self more generally, and to orient the self in space and time. The result of these impacts was to undermine the sensemaking ability of the department, in terms of making sense of the immediate environment as well making sense of the department's response to it.

Regardless of the negative publicity surrounding the police department's performance after the hurricane, a careful consideration of the entirety of the situation suggests that, in some ways, the department performed remarkably well. Despite the department's prior problems with corruption and low morale, despite a lack of planning by the city, despite having lost access to much of the infrastructure of police work, despite being forced to work in a filthy environment, despite having no homes to return to, despite being unable to lock up criminals in jail—the majority of the police force stayed in place and in many cases heroically improvised solutions to almost impossible problems.

This is not to say the department's performance is beyond criticism. But it does suggest certain resources that ought to be in place if other local emergency response agencies are to avoid these problems when faced with similar circumstances. Better, more comprehensive planning and prepositioning of resources for use in extreme events would certainly help. This might need to include resources that are commonly thought of as beyond the scope of emergency response organizations, like housing, food, and personal hygiene facilities for responders. Improvisation is an ability that can be learned and improved upon with practice, as any New Orleans jazz musician can tell you. Emergency response exercises could stress improvisational skills to a much greater degree by routinely taking away familiar infrastructural resources and forcing responders to develop work-arounds. Finally, disaster response plans could be more realistic if they took into account the likelihood that, in an extreme event like Katrina, some local emergency response agencies could be as severely affected as any population in the disaster zone, and might even be unable to effectively request help. More research is needed to determine whether these kinds of interventions would be effective, or if others would be more effective, at bolstering the ability of emergency responders to respond effectively to a disaster affecting their local area.

CHAPTER 4

Power Loss or Blackout:
The Electricity Network Collapse of August 2003 in North America

Timothy W. Luke

On Thursday, August 14, 2003, the worst and most widespread loss of electricity during a blackout in U.S. history occurred at 10 minutes and 48 seconds after 4:00 p.m., knocking out electrical power in eight states as well as large areas of Ontario and Quebec in Canada. The outage happened almost simultaneously. At first, it was thought to be the result of a terrorist attack, because of horrendous destruction so recently experienced in New York and Washington, DC on 9/11. Soon, however, this electricity network failure was discovered to have been merely an accident, although it was not known for weeks exactly how, where, or why the accident happened. Former Secretary of Energy in the Clinton Administration, Governor Bill Richardson of New Mexico, was reported the next day on the front page of *The New York Times* as observing that "We are a major superpower with a third-world electrical grid. Our grid is antiquated. It needs serious modernization."[237]

A handful of smaller municipal and regional utilities allowed a few urban areas in the Northeast to enjoy uninterrupted service during this massive lapse of service, but such widespread electrical power loss disrupted many cities and essentially created gridlock in New York City, Albany, Syracuse, Buffalo, Rochester, Erie, Cleveland, Detroit, Toronto, Ottawa, and hundreds of other smaller cities and towns. Thirty million people lost power in the famous 1965 blackout, while nearly 56 million experienced the power outage of 2003 in hundreds of cities and towns.[238] Beyond the massive chaos, the blackout cost $6 to $10 billion in lost output, wages, sales, and earnings (Figure 4.1).[239] The simultaneous complete loss of power by such a large population in the world's allegedly sole remaining superpower is an event worthy of careful further consideration to consider how vulnerable to electrical blackouts such populations truly are as well as to gauge why network failures can be understood as ever-present structural threats to cities embedded in the contemporary workings of modern markets and technics as such.

The underregulation of the American electrical industry, and disinvestment in the technical infrastructures of the electricity grid, follow from the neoliberal consensus emerging triumphant all across the United States in its commercial, financial, and monetary centers. This attitude follows both "the pauperization of the state" and "the commodification of public goods," and this fragility of today's electrical grid comes from liberal democratic governments' all too "willing submission to the financial markets."[240] In all cases, this capitulation is leading to the gradual degradation or a rapid meltdown in vital public services, as the August 2003 blackout shows.

Figure 4.1 People take to New York's streets during the 2003 blackout. Source http://commons.wikimedia.org/wiki/File:2003_New_York_City_blackout.jpg

THE POINT OF DEPARTURE

In reassessing this severe electrical infrastructure crash, one must resist the alternating currents of operational stability, service certainty, and technical normality routinely expected from urban systems. It is vital instead to reroute notions about the politics and economics of electricity grids through new conceptual transformers to bring criticism to peak output. The massive power outage in North America on August 14, 2003 is an excellent case in point for taking these turns. This event shows that many analytical tools in science and technology studies are not adequate to the tasks of interpreting today's built environment. In fact, existing conceptual frameworks often occlude what needs to be analyzed, who needs to be criticized, and what must be done to overcome embedded trends toward powerlessness and inequality.

To anchor such claims, Jameson's observation about "the postmodern condition" should be the point of departure. That is, it is what exists "when the modernization process is complete and nature is gone for good. It is a more fully human world than the older one, but one in which 'culture' has become a veritable 'second nature.'"[241] Ironically, however, as technoscience becomes a second nature, it scrambles together human subjects and nonhuman objects as quasi-subjects and quasi-objects, working together collectively in systems of systems. This human world rests simultaneously upon the creation, maintenance, and suppression of a parallel nonhuman world. As Latour suggests:

Modernity is often defined in terms of humanism, either as a way of saluting the birth of "man," or as a way of announcing his death. But this habit itself is modern, because it remains asymmetrical. It overlooks the simultaneous birth of "nonhumanity"—things, or objects, or beasts—and the equally strange beginning of a crossed-out God, relegated to the sidelines.[242]

These correlative and collaborative domains of nonhumanity usually are labeled as "the environment"—both built and unbuilt. A fine example is the national electricity grid. Concealed in the center of the action with "modern man," the grid combines things, places, systems, and spaces that empower "man" for "modernity." Like the society they serve, these networks constitute and empower many quite modern human inequalities in their skewed structures of social opportunity, as these forces immerse almost everybody and everything on the grid in their workings.

Today's electrical grid, then, is an intriguing site from which to observe Latour's generation of new Nature/Society/God constructs in checked-and-balanced pairings of transcendence and immanence. Simultaneously, as they flip on the lights, those who are modernizing or modernized can believe:

> They have not made Nature; they make Society; they make Nature; they have not made Society; they have not made either, God has made everything; God has made nothing; they have made everything.... By playing three times in a row on the same alienation between transcendence and immanence, we moderns can mobilize Nature, objectify the social, and feel the spiritual presence of God, even while firmly maintaining that Nature escapes us, that Society is our own work, and that God no longer intervenes.[243]

Of course, massive blackouts move many to contradict Latour with claims that such temporary power outages are "an act of God." Nonetheless, these constitutional principles permit technified hybrid collectives, or "the association of humans and nonhumans,"[244] to proliferate—as both abstractions and instrumentalities typically labeled as science, technology, culture, society, or markets—on the terrains of modernity, especially when "the power is on," as ordinary urban sprawl.

When "the power is off" one sees how much these hybrids are the fabric of our lives, those good things corporations bring to life, or where science and technology get down to business. Meanwhile, humans ignore the hybrids, denying the very existence of such hybridized actualities by clinging resolutely to conventional Enlightenment fables of how live human subjects always control and dominate dead nonhuman objects through science and technology.[245] As Latour maintains, such fables usually render "the work of mediation that assembles hybrids invisible, unthinkable, unrepresentable"[246] in undisrupted cities.

Still, as the United States–Canada Power Outage Task Force hints, the grid is one place to seek and find the powers of this hybridity. Obliquely gesturing toward both the human and nonhuman elements, the Task Force observed how North America's electricity system is a magnificent engineering project over 100 years in the making: "This electricity infrastructure represents more than $1 trillion (U.S.) in asset value, more than 200,000 miles...of transmission lines operating at 230,000 volts and greater, 950,000 megawatts of generating capability, and nearly 3,500 utility organizations serving well over 100 million customers and 283 million people."[247]

Such modernities rest on two counterpoised ontopolitical rails: *translation*, which creates mixtures between entirely new types of beings, "hybrids of nature and culture"; and *purification*, which denies the mixtures of translation as it "creates two entirely different ones: that of human beings on the one hand; that of nonhumans on the other."[248] As long as these practices are treated as divided and different, then people can think and act "modern." Translation ties together conventionalized constructs in networks of quasi-subjects/quasi-objects between "a natural world that has always been there, a society with predictable and stable interests and stakes, and a discourse that is independent of both reference and society."[249] The analysis of modernity, then, should confront these "quasi-fication" processes in hybridization, because clearly "objects are not the shapeless recreatables of social categories neither the 'hard' ones nor the 'soft' ones.... Society is neither that strong nor that weak; objects are neither that weak nor that fabricated, much more collective than the 'hard' parts of nature, but they are in no way the arbitrary receptacles of a full-fledged society."[250] Big technological systems, like electricity companies, power lines, and generating plants, cannot be understood fully without seeing how these quasi-fications constantly are pulled into the purifications of policy discourse or fail to disclose their technical–economic translations in action.

In *The Informational City*, Manuel Castells spins a vision for the new order of information society—one marked by "the space of flows."[251] While his observation is not necessarily wrong, it clearly is not remarkably new, strikingly insightful, or even essentially right. Any close reading of the second, and even the first, Industrial Revolution shows that the mechanization, electrification, and the motorization of society in the eighteenth and nineteenth century had created already their own mobilized "space of flows" for decades.[252] With electricity's embedded energy, if not embedded intelligence, and the "ubiquitous motoring" it made possible from the 1880s through the 1950s, electrification agencies anticipated many of today's dreams of "ubiquitous computing." Consequently, as the grids grew, many social practices were reshaped radically by electricity as buzzing assembly lines, humming telephone lines, and crackling electrical lines became fixtures in the ordinary background of industrial society. Thus, despite decades of energy-saving

efforts in the United States, in 2004 the nation consumed 167 percent more electricity than it did in 1970, and electricity's share of total energy usage rose from 25 to 40 percent during this same period.[253]

This massive electrification of contemporary society was, in many ways, as profound a foundational change in the social order as today's computer-driven informationalization.[254] The electrified city, factory, home, and marketplace in the United States or Canada are predicated upon organizing activity and structure around a commodity which must be generated continuously at base load levels, pushed up at intermittent spikes to satisfy peak use surges, moved constantly to sustain service, linked seamlessly in a grid of grids to assure redundant back-up capacity. Yet, it also ironically cannot easily be stored, sold outside of the grid, or substantively transformed as a commodity without its unique transmission networks and generating stations—all of which are charged with power.

The flows of electrical energy, then, have shaped the spaces and sites of urban–industrial life very significantly since Thomas Edison opened the first municipal electricity plant on Pearl Street in Manhattan during September 4, 1882.[255] Flows and forms already came together in the Gilded Age to both structure agency, and activate structures, with constellations of power in those countries that could deploy such powerful systems of empowering systems—water, sewer, gas, electricity, telegraph, telephone, road, and rail—to organize the conduct of their subjects and program such objects with conduct.[256] Nation-building basically entwines itself with machination-building as the hybridization of machinic systems, human populations, and territorial spaces.

POWER LOSS: FADE TO BLACK

The attainment of popular sovereignty, or self-rule, during and after the Enlightenment clearly constitutes a major milestone in Western civilization. Nonetheless, it is the empowerment of people through electrification with its own media of control, information, and order in the nineteenth and twentieth centuries that many appreciate far more. Indeed, "getting power" on the grid makes the modes of individual agency and social structure set forth by the Enlightenment more materially possible.[257] This transformation is not juridico-political as much as it is techno-economic. Only a few studies by political theorists or cultural critics have investigated any of its ramifications. While popular sovereignty plainly marks a transfer of authority to the people, it is getting "power service to the people" through electrical energy via all of the motors, lights, systems, and devices—now considered the sine qua non of "modernity"—that must be more systematically reviewed as an ontopolitical condition.[258] Societies can gain popular sovereignty, but have no electrification, while other societies attain electrification without enjoying popular sovereignty. How does this unevenness develop?

Charles Perrow in *Normal Accidents: Living with High-Risk Technologies* (1984) makes many observations about why new, unfixed technologies inherently involve high risk in their use, but he also lays much of their risk at the doorstep of their systemicity. That is, as the system-building process unfolds we create systems—organizations, and organizations of organizations—that increase risk for the operators, passengers, innocent bystanders, and for future generations.[259]

Most importantly, there is the ordinary actuality of the operations as risk-ridden enterprises which carry within their daily operational practices a "catastrophic potential," and, yet "every year there are more such systems."[260] Normal accidents occur simply because of the conjunctures of contingency needed by the system to work. Occasionally, work fails, exceeds, lags, or otherwise deviates from ordinary parameters just enough,

it morphs from acceptable ordinary actuality into unacceptable catastrophic potentiality.

The "normal accident," then, is so arresting and awful because it reveals the more liberating and awesome accidental normality of system-building also unfolding behind the machination. Systems are built, but their construction is necessarily fitful, irrational, unsystematic, and improvisational. Behind manifestly ordinary urbanization, extraordinary urban disruption latently awaits some triggering event. These systems' normality—for all of their apparent elegance, power, or order—remain as accidental as they are intentional. Most importantly, this accidental normality cannot escape the eruption of normal accidents (Figure 4.2). One only can, as Perrow argues, calculate their probabilities, manage their risks, and contain their damage to the best of one's systemic, albeit often equally improvisational and ready-made, abilities.

Once electrified, people frequently lose power all of the time. While they might retain popular sovereignty or even endure without it, no one relishes blackouts. Such "loss of power," "power failures," "power outage," or "power interruptions" as moments of tremendous disempowerment, therefore, need to be monitored. Without power, there is a new vulnerability in everyday life to persistent systemic gridlock or even rapid demodernization as "power loss" deeply disrupts the ordinary motorized, mobilized, mechanized routines of urban societies and economies.[261]

What are some of the implications of these network failures with regard to their impact on social control, informational operation, and political order? The frequency of such events of failure, outage, or loss has led many to see only the "normal accidents." Yet, it seems clear that "normal accidents" essentially rest upon, and then intermittently reveal catastrophically through loss, this deeply embedded fragility, or the "accidental normality" that has been hastily constructed, rapidly generalized, and then quickly natu-

Figure 4.2 Commuters walk home across the Brooklyn Bridge in New York City half an hour after the start of the August 14, 2003 blackout. Source: Photograph by William Hall http://www.panoramio.com/photo/161808

ralized as modernity. At this point of near complete nonawareness of fragility in the workings of complex social systems powered up by electricity, unacceptable catastrophic potential can just happen.

The accidental normality of electricity networks in the United States can be seen in both its regulated past and its deregulated present as well as the incredible pace at which it works. Power service today rests, in large part, upon this pastiche of corporate power generating productive capacities behind modern subjectivity, but all these circuits can black out with any power outage. Power losses vary as events. During 1977, widespread violent looting occurred in New York City as thousands of people ran riot for blocks and blocks, stealing TVs, food, refrigerators, and furniture. In 2003, most people were calm; they slept in darkened streets, or walked for miles to powerless homes.[262] So even the experiences of normal accidents themselves can give rise to expedient constructions of them as an accidental abnormality or abnormal normality.

The electrical grid that powers the United States and Canada is immense, encompassing over 235,000 miles of lines whose wires and cables carry ordinarily anywhere from 110 volts to over 230,000 volts.[263] Because it is vital for the nation's economy, the grid also constantly grows more unstable and unpredictable as a complex system. By 2003, the United States was generating and using 710,000 megawatts of electricity over the grid.[264] From 1993 to 2003, the demand for electrical power rose 35 percent; at the same time, deregulation turned the grid into more mangled conjunctions of markets that responded to this demand from the growing economy. Electricity itself became more easily commoditized, like copper, oil, or pork bellies, as the grid grew in scope and density. Moving energy over the grid in wholesale monetized transactions increased 400 percent from 1993 to 2003, but new investments in electricity infrastructure only rose by 18 percent.[265] While it is huge, essential, and profitable, the grid itself also has proven to be a site of growing disinvestment with tremendous vulnerability.[265] Because underinvestment, misunderstanding, and indecision have become institutionalized as features of its daily operations, the grid's deregulated conditions of operation create the potential for tremendous crises to cascade out of its now supercharged, overburdened, and undercapitalized networks.[266]

The complex brittleness of the North American electrical grid parallels that of world financial markets. The massive financial meltdown of 2008 essentially is prefigured by the power outages of 2003. Investment in the grid's transmission infrastructure has not kept pace with the rapidly expanding practices of new wholesale and retail electricity markets. The neoliberal deregulation of electricity generation and consumption permits utilities to build more power plants, consumers to select electricity sources, while compelling companies to figure out how to access this new generating capacity, meet specific consumer demands, and manage base and peak power delivery in this environment.

Strangely enough, these markets usually afford fresh profits from flexibly fabricated generating capacity and consumer demand. Nonetheless, they rely upon a rigidly regulated grid that is much more constrained, was not designed to carry power so flexibly, this quickly, or so rapidly by linking generating plants and paying customers increasingly separated by greater distances.[267] Since there is more profit to be had from generating capacity, companies have no incentive to invest in their grids, and regulators do not have sufficient authority to compel them to make those investments (since their regulating mission is pitched at far smaller, more local, state-centered and more vertically integrated electricity producers).

BLACKED OUT: POWER LOST

The massive "power loss" event of August 14, 2003 was presaged by some minor power supply abnormalities in First Energy Corporation's service area in northern and eastern Ohio during the afternoon of August 14. These fluctuations then became severe around 3:05 p.m. To meet high summer demand for power, First Energy was wheeling electricity through its lines to meet the needs for power in Ontario. First Energy, however, had no emergency blackout plans in place, so it could not shut off any adversely affected zones in the grid quickly enough to avoid a rapid grid collapse. Had First Energy only blackened out the metropolitan region around Akron and Cleveland by shedding up locally to 1,500 megawatts that day, the systemic collapse throughout the Northeast probably would have been avoided. Instead, large flows of electricity that were running across the Great Lakes through First Energy's lines suddenly reversed. Due to these unstable load conditions in northern Ohio, current rushed back into the grid's networks, overloading power lines and tripping utility circuit breakers.[268]

As Michael R. Gent, the President and Chief Executive of the North American Electric Reliability Council remarked, "This whole event was essentially a 9 second event, maybe 10."[269] Nonetheless, during that 10 second window of instability, dozens of power lines, nearly 100 generating facilities, and almost 62,000 megawatts of power capacity went down, essentially before any human beings could accurately detect, assess, and then react to the power loss. Over 50 million people lost power for minutes, hours, or days and billions of dollars in economic losses were incurred due to an odd twist in a 9 to 10 second cusp of chaos. Yet, this chaos was an artifact of the market itself, which concocts its own turbulent cross-currents out of speed and power.[270] Capacity, however, implies incapacity, and those twin vulnerabilities are engineered all the way down into the connected systems powering most of northeastern Canada and the United States.

At the point of collapse, First Energy

Corporation had made small mistakes that snowballed into a major emergency. Pathetic but true, it had been routinely neglecting its own corporate policies about tree trimming near power lines (Figure 3.3). One of its generating plants failed at about 1:30 p.m. due to technical problems, but this small loss should have been manageable with the right response. Due to the hot day and tremendous power load, however, the lines dropped into high trees near the transmission wires, so three major lines shut down between 2:00 and 2:15 p.m. within First Energy's grid.[271] However, the central coordinating agency for power in the region, MISO (Midwest Independent Transmission System Operator) by happenstance had its "state estimator" computers off-line due to a small malfunction, an incomplete repair, and then a full switch-off during the repair technician's lunch hour. Thus, MISO and First Energy were monitoring the grid on August 14 with incorrect and inadequate power load data. When the power lines failed, MISO could not detect the aber-

ration. First Energy also could not adjust its own power management programs. Transmission problems that neither entity was aware of, or able to quickly manage, then triggered the collapse between 3:00 and 4:00 p.m. as First Energy's own computer alarm system for system disruption also failed.[272]

By 3:30 p.m., as First Energy's lines were going out of service due to these line loading and tree-brushing coincidences, its control rooms were unaware of the failures, MISO's "state estimator" computer program had not signaled any operators about the aberrations. Had First Energy cut off Akron-Cleveland around 4:00 p.m., the blackout plainly, according to Alison Silverstein, a Federal Energy Regulatory Commission Advisor, "would have been a local Ohio problem" (see Figure 3.4).[273] It failed to make this decision, and a crush of kilowatts cascaded unpredictably, and almost simultaneously, as a massive outage across parts of Ohio, Michigan, Indiana, Pennsylvania, New York, New Jersey, Connecticut, Vermont, and Ontario.

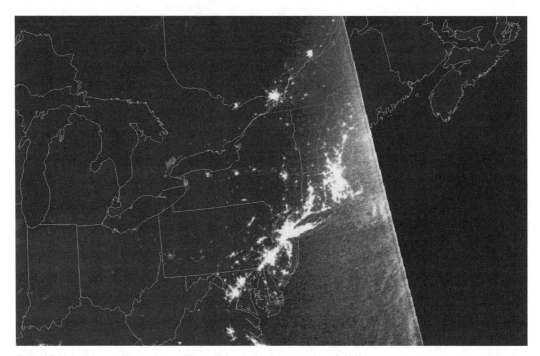

Figure 4.3 Night Lights satellite image about 20 hours before the blackout taken Aug. 13, 2003.

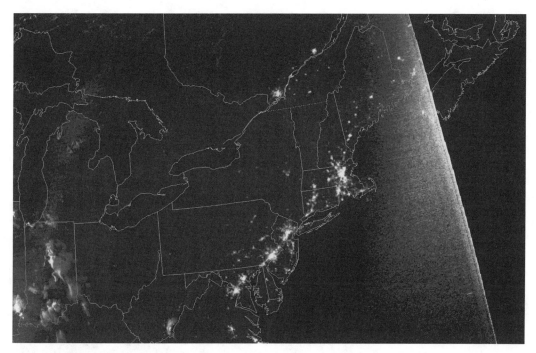

Figure 4.4 Night Lights satellite image about seven hours after the blackout.

In New York City alone, "the City Council of Finance Office estimated the blackout cost the city up to $750 million in lost revenue, up to $40 million in lost tax revenue, and up to $10 million in overtime pay for the first 24 hours after electricity went out."[274] These losses were compounded city by city, million for million all across the region affected by the blackout. Early estimates of a $1 billion total loss in the nation's $11 trillion macroeconomy saw the blackout as "a nuisance" or "rounding factor" sort of loss,[275] but the losses obviously were much more significant in the microeconomies of the many small businesses that deal in perishables, local governments that spent extra funds on relief worker overtime, or utility companies forced to cycle their blackened out grids back to full service. Moreover, the accidental normality of such "normal accidents," like this August 14, 2003 network failure, remains fully embedded in place for the next concatenation of comparable coincidences.[276] The accidental architectures of the grid as a system as

well as the intentional gambling of regional power markets to maximize profits create severe personal and collective vulnerability by all utilities simply lighting up their wires.[277]

Capitalism rather than terrorism, then, is to blame for events like the August 14, 2003 Northeast blackout or the 2001 California electricity shortages. Deregulating power generation in 1992 has enabled a new, mostly uncontrolled market in wholesale electricity to develop, but this electricity is still transmitted over high-tension lines and a distribution system that is locally regulated to protect retail business and household users. Not surprisingly, most companies have invested more in generating new electricity while they have neglected investments in power lines, transformer yards, or grid controls. More and more electricity, in turn, is moving over a system of systems that was designed to carry lower loads, to receive better maintenance, and to serve smaller regional markets. Old infrastructures and institutions—built to work under closer regulation and for more local users—now are forced to serve with less regulation and more national (or international) customers. Again, these outages suggest the accidentally normal quasi-polis of people and things—working together at different times as quasi-objects and quasi-subjects—is failing to sustain the hybridized commercial regime that brings these people and things together transnationally to operate in this manner.

Conventional social science focuses with realist juridico-legal categories upon people and their quest for power within each national polis as well as the will to dominate of every polis over others. An equally realistic reading of these times must look at the techno-economic quasi-polis of people and things coexisting in international, national, regional, and local systems in which "all that is solid melts into air, all that is holy is profaned, and man is at last compelled to face with sober senses, his real conditions of life, and his relations with his kind."[278] There are different struggles among men, and women, within the quasi-polis over how to first make possible, and then take to relying upon these probable nonhuman conditions of life with their own kind and other machinic systems. Living amidst the accidental normality of any built environment today is made possible, or impossible, by the power and knowledge embedded in a baroque material regime required to run the water, gas, sewer, road, telephone, radio, television, and electricity systems interwoven into every quasi-polis against ontopolitical others tied to different machinic alternatives.

Power companies, as they operate their systems of systems in global markets, are complex capitalist machines fabricated by accident and design to produce certain goods and services.[279] The seat of empowerment, understood as the generation of development, modernization, or even civilization, now flows through an accidental normality resting upon the inherent risks of such quasi-political systems.[280] Inasmuch as the United States and Canada depend upon innumerable corporate acts and company artifacts shaped by particular enterprises in specific settings, such accidentally normalizing infrastructural powers, especially electricity systems, now push most ideas and material things to be mobilized for the advance profit-seeking corporate strategies. Before powering society up, their industrial ecologies of development and modernization are hard to envision.[281] In the twenty-first century, however, power outages are gaps of demodernized, undeveloped existence.

Electrification's empowerment displays how market-based technologies of production and the self cogenerate new linkages between objective systemic productivity and subjective idiosyncratic consumption for producers and consumers in the social regimens of electrified life. The consumers of corporate commodities are redesignated through their purchase of commodities to play the role of capital asset, causing "the ultimate realization of the private individual as a productive force. The system of needs must wring liberty and pleasure from him as so many functional elements of the reproduction of the system of production and the relations of power that sanction it."[282] Corporate plans for social transformation gain life, liberty, and property through the buying decisions of individuals rather than the other way around. For transnational businesses, the liberation of personal "wants" or individual "needs," as they are allegedly felt by everyone anywhere, is fixed by making more and more commodities hitherto inaccessible in many markets available to all who desire them.

Power outages "out" how power now works in this era, because blackouts briefly fail to "blacken out" the hybridizing links to quasi-subjectivity in these objective commodity chains. Generalizing the availability of electricity, and then finding "needs" for it through new appliances and artifacts to sustain system load is a perfect case in point.

Consumer goods can be supplied once new subjects are recognized as having the demand functions expected from being "good consumers," who must then get their "good things." Subjectivity is redefined through electrification as a material need for coexisting with electrical goods, and modern subjects are those who can be defined by their material demand for electrified goods and services designed to supply and thereby satisfy them. Powered-up objectivities, in turn, shape empowered subjectivity through this machinic order. Here Baudrillard observes:

> "The *consumption* of individuals mediates the *productivity* of corporate capital; it becomes a productive force required by the functioning of the system itself, by its process of reproduction and survival. In other words, there are these kinds of needs because the system of corporate production needs them.... It is *all* capital."[283]

On the grid, the elective affinities of corporate expertise and authority bring technologies of the self (consumer decisions) together with technologies of production (producers adding value) in the world's industrial ecologies.[284] Without the various electric toasters, dryers, radios, stoves, coffee makers, computers, lawnmowers, or televisions, there is no base load for operating systems of electricity.

Ideologies of competitive corporate growth realized through the exploitation of labor are inscribed in each electrified commodity, even though these powered-up objects are delivered to submissive consumers as true tokens of new empowered identity found from their collective liberation by electrification. Any "power loss" drains the requisite flows of energy needed to order this quasipolitan system around its routines of machinic regulation. The longer electricity is lost during blackouts, gridlocks, or outages, the more demodernization stares out of every dead power socket and inert appliance.[285]

Expertise and ownership construct a program of command and control, and they communicate themselves through the ever shifting normalization routines of electrical commodities. When consumers admit that "they're living it," or that products give them "that feeling," or that buying the right stuff gets them "connected," it is clear that individual subjects have become repositioned by their possessions in the manifold agendas of transnational globalism. General Electric has prided itself in "bringing good things to life," and electrification is the quasi-political foundation of how those good things are brought into living. As Foucault notes, "individuals are vehicles of power, not its points of articulation."[286] Electrification—either when it is online or as it slips away in power outages—shows how commodities work as effective relays of corporate management. Inasmuch as their generic capacities for market-mediated individualism become materially articulated in their specific effects over the screens of power upon one, some, many subjects as well as collectivized as the universal affects within one, some, many objects, the grid underpins clusters of commodified compliance exerted by the appliances of electrical modernity.[287]

Fleeting commodities, like electrical power, as their prices rise and fall in the markets that the transnational corporate firms produce, can operate as "a polymorphous disciplinary mechanism" for corporate power.[288] Individually and collectively, the machinic assemblies producing these artifacts carefully have cultivated over the past century "their own discourse," and "they engender...apparatuses of knowledge (*savoir*) and a multiplicity of new domains of understanding."[289] Within electrified systems of systems, powered-up electrical services are simultaneously carriers of discourse, circuits of normalization, and conduits of discipline, which firms utilize to possess their individual proprietors with the properties of their systems as reified as artifacts of personal property.

Maybe commercial artifacts circulating on the grids of corporate commodification should be explained as the conducting media for the conduct of *"a society of normalization"*[290] in which adding value to the artifacts of commerce or purchasing power over consumer goods become the sine qua non of being a good consumer with normal tastes and attitudes. To live on the grid is "enjoying the good life," while leaving the grid is an antisocial act that expresses the intent to retreat into premodern darkness and drudgery. Of course, normality is never univocal or monodirectional; it is multivocal and polydirectional, as any strategic analysis of the corporate authorities intent upon exploiting global purchasing power in electrified markets soon reveals.[291]

From Edison's first push in New York through today's Internet explosion, electrification is a perfect case in point for these changes. Capitalist exchange, under these conditions, brings an electrifying subjectivity of object-centeredness. Everyone becomes what they buy, everyone buys what they are, have been, and will be. "In the end," as the architectures in these systems of systems prove, developed societies and modernized peoples become "destined to a certain mode of living or dying, as a function of the true discourses which are the bearers of the specific effects of power."[292] Electricity, gas, water, telephone, or television services are now a conventionalized script and package of "civilized life" for any civic group, and all quasi-political imperatives require people to acquire these goods and qualities in close coevolutionary collaboration with private enterprises with national economies operating as machinic systems.[293] Without the quasi-political technics of urban life, there would be no polis. Without this quasi-polis, there is no politics.

As the focus of power and locus of subjectivity in our world markets, the quasi-political system of systems ironically is being built. It is already always accessible, mostly within the apparently almost accidental anarchy of markets. One view of the many accidents "that gave birth to those things that contin-

ue to exist and value for us" can be found in contemporary realist celebrations of the environment-generating forces at work in today's world marketplace.[294] Like Smith, this technocultural reading of quasi-politics also must trace out "the productive powers of labor, and the order, according to which its produce is naturally distributed among the different ranks and conditions of men in the society."[295]

The principles that now govern a whole system of systems as well as the particular modes of labor that maldistribute different classes and nations within these systems do not cluster in a clutch of accidental accords.[296] Instead there seems to be an imminent design for inequality in access, power, status, and wealth here, even though their final forms remain obscured in the unfinalized empiricities of modernity. Fortunately, quasi-politics are one of the more complete registers for disclosing these designs as they mediate both power and knowledge in the concrete manifestations of things in the ecologies of everyday life.[297]

Beneath highly planned exercises of rationalization in the quasi-polity of technified lifeworlds, consumers are left living with consequences far beyond anyone's command, control, communication, or intelligence[298] and living on "the grid" underscores this point. The allegedly growing calculability of instrumental rationality actually brings new measures of incalculability—unintended and unanticipated—rooted in instrumental irrationality along with it as planned environments create unplanned ones, as affluent environments degrade poor ones, and as machinic environments contaminate organic ones in the electrified regimes of accidental normality. As the financial crises of 2008 suggest, vulnerability now is the material basis of modernity. Indeed, the "high standards of living" afforded to most in what can be described as America's "actually existing industrialism" all rest upon increasingly failure-prone, failing, or failed systems. Astounding events at the outer limits of failure, like the power outages of 1965, 1977, and 2003 reveal this fact in its rawest form, because the accidental normality of many other infrastructural systems holds equally disruptive potentials for "power loss."

CONCLUSIONS: GRID-LOCKED

To anticipate the incidence of "normal accidents" in large complex systems, as many risk assessment exercises have done, is, at the same time, to participate in the generation, and then naturalization, of an "accidental normality" at the core of systematized complexity. The intrinsic physical, operational, mechanical, chemical, or biological qualities of many technologies direct engineering down particular paths of creation; yet, in many other ways, there are always contingent cultural, economic, political, or social choices that must be made at peculiar points of decision making whose boundaries are set more by aesthetics, cost, power, or status. Such outcomes basically are accidental, but their attainment is normalized as frequently and easily as mathematical constants or biophysical regularities. Behind their normalized constancy, contingent accidental irregularities congeal only as long as the same cultural, economic, political, or social conditions hold.

From perspectives tied to "deep technology," this view of the electrical grid as a semiautonomous mystery in the built environments was underscored by many assessments of what August 14, 2003 meant. The physicist Phillip F. Schewe suggested that the grid's structure alone makes it likely to collapse: "just by sheer complexity, adding one thing to another, larger and larger disturbances are going to occur."[299] Technology critic Edward Tenner remarked, "the grid definitely makes life safer and more reliable," but the complexity that Schewe cites will lead to collapses to a point where "the dominoes can start to fall over a wider and wider area."[300] Along with

Perrow, these experts argue "the incomprehensible complexity of the grid comes with its own irreducible pathologies," such that "the grid—complicated beyond full understanding, even by experts—lives and occasionally dies by its own mysterious rules."[301]

In this transnational quasi-polis, as millions of Americans and Canadians discovered in the blackout, ordinary processes of democratic legitimation fail when "the system does not deliver." Modern industrial revolutions with their many products and by-products are highly invested in both mysterious rules and irreducible pathologies simply to sustain their economic development. Any formation like the electrical grid always "remains shielded from the demands of democratic legitimation by its own character" inasmuch as "it is *neither politics nor non-politics,* but a third entity: economically guided action in pursuit of interests."[302] Because of property rights and expert prerogatives, most occupants of this increasingly planetary quasi-polis have yet to realize fully how "the structuring of the future takes place indirectly and unrecognizably in research laboratories and executive suites, not in parliament or in political parties. Everyone else—even the most responsible and best informed people in politics and science—more or less lives off the crumbs of information that fall from the tables of technological sub-politics."[303] In turn, clear understandings of these machinational world systems and their quasi-politan relations become a new means to organize, or cyberorganize, "the conduct of conduct" as another governmentality circuit for citizens and consumers.[304]

The far-flung blackout of August 14, 2003, like those during 1965 and 1977 in the same area, or the California blackouts of 2001, essentially had to be reinterpreted as an unfortunate coincidence, an uncanny happenstance, or a pure fluke.[305] Others, however, highlighted how strangely comparable this incident for each disrupted city was to intentional acts of terrorism, cyberwarfare strikes, or planned incidents of sabotage (see Graham, this volume). Power loss on this magnitude, to play off of the bizarre design parameters of neutron bombs that kill people but leave most built structures essentially intact, is an "electron bomb" which totally paralyzes built structures but thoroughly threatens the people dependent upon them. Disrupting the flow of electrons in network failures makes the quasi-political system of systems inert, inoperative, or even inimical for the humans and nonhumans that rely upon them.

Humanist visions of power fail us when considering "power outages" or "getting power" on the grid. Becoming invested in progress, one loses the original promise of liberal rational autonomy from the polis as the powers of the quasi-political systems prove to be often illiberal, mystifying, and enslaving. While all are continuously promised greater access, education, and knowledge through technologies, few really know how everyday life "gets done" technologically. And even though the opportunity for control and cooperation is promised, few truly own how anything is realized. Modernity has become a means whereby the few who know-how and own-how maintain domination over the many who do not know-how or own-how, even though this power outage is hidden in the brightness and buzz of electrification. The illusion of progress through greater education and broader opportunity, in fact, always belies grittier realities of an almost complete "power loss" in the exploitative avarice fostered by growing disinformation and greater dispossession.[306]

Consequently, it is the nonhuman politics of quasi-politan inequity that need to be more closely demystified by humans in their policy, theory, and practice to address today's economic crises and political contradictions. Here, the power loss and power outages from blackouts on the grid underline the unwillingness of most contemporary humans to come to terms with where they really are and what they actually can do, namely, always exist seconds away from massive power outages in disrupted cities.[307] Only when they are left in the dark by blackouts do their baseline existential conditions of near complete power loss become glaringly obvious.

CHAPTER 5

Containing Insecurity:
Logistic Space, U.S. Port Cities, and the "War on Terror"

Deborah Cowen

"The utopian vision of a network without hierarchies is an illusion, an attractive theory that has never been implemented except as ideology."[308]

In February 2007, Dubai World Central in the United Arab Emirates (UAE)—the largest master-planned settlement on earth, the "most strategically important infrastructure development"—opened the gates to one of its most ambitious projects: Logistics City (Figure 5.1). Dubai World Central encompasses a wide range of specialized zones and uses, but Logistics City is perhaps the most unusual and intriguing element of the plan. Logistics, the design and management of supply chains, has become so critical to just-in-time globalized production and circulation systems over the past four decades that there is now a city named in its honor. Urban space is conceived and produced for the singular purpose of securing the management and mass movement of globally bound *stuff*.

Logistics City is already setting standards that are reverberating along transnational supply chains. It sets momentous precedents for the production of urban space and the politics of infrastructure protection, reaching far beyond Dubai and the Gulf region all the way to the ports of the United States. United Arab Emirates officials claim that Logistics City will be "the world's first truly integrated logistics platform with all transport modes,

logistics and value added services, including light manufacturing and assembly, in a single customs bonded and Free Zone environment." The plan unites the Port of Jebel Ali—the world's largest human-made harbor and the largest port in the Middle East, with the new Dubai World Central International Airport, planned to have the world's greatest capacity upon completion. Logistics City entices firms to become tenants with offers of fifty-year no-tax guarantees, no caps on capital inflows or outflows, and no labor restrictions. It is part of an aggressive effort to diversify the Emirati economy beyond oil into alternative growth industries.[309] Dubai Logistics City prides itself for offering the "latest technology solutions" not only for transport and customer service, but for "security" as well.[310]

Concern for security is significant. The term is rarely defined and yet "security" is invoked as a self-evident priority in urban development and infrastructure projects around the world. However, it is the security of *supply chains* rather than the people who live and work in the city that is at the core of a logistics lens. Order and efficiency guide the design of built form and govern the distribution of uses and users in space, as the plan below conveys. Yet, while logistics is ostensibly about efficient movement and undisrupted flow, the plans for Dubai's Logistics City reveal that flow is achieved through a

Figure 5.1 Dubai Logistics City, Source: Brian McMorrow 2007.

proliferation of new borders and sociospatial ordering and control. Some forms are clearly visible in the landscape, while others are hidden from immediate view. On the one hand, the entire vision for the city resembles a computer motherboard; an engineered environment without chaos, disorder, or detritus, let alone signs of life. Dubai Logistics City is ordered electronically through biometric access cards, security gating, cameras, and other technologies. Yet, the most significant form of spatial containment is likely the "Labor Village." Officials promote the "village" as the "provision of integrated blue collar housing with full range of facilities." But despite the clean design and lavish landscaping, artists' renderings still conjure an air of prison architecture. Described by local media as a "luxury labor camp,"[311] Labor Village will eventually occupy 14 million square feet of land and hold 87,500 beds.[312]

Labor Village is a response to the grow-ing number of labor actions in Dubai, which create increasingly frequent disruptions to commodity flows. But if Labor Village is a carrot dangled in attempt to buy workers' loyalty, official response to labor activism has also taken the form of a big stick. Strikes and trade unions remain illegal in the UAE and participation in either can result in permanent expulsion from the country.[313] This is one stark indication of the extreme conditions operative in Dubai; the vast majority of workers are on temporary work permits without any formal citizenship. While countries like the United States rely heavily on noncitizen labor, no place on earth matches the UAE in this regard. Noncitizens make up 99 percent of the private work force (two thirds of which is South Asian) and there are more than 4.5 million residents but only 800,000 Emirati citizens.[314]

A paradox with transnational salience lies at the core of Dubai Logistics City. Insuring flow to commodity flows seems to require containment. On the one hand, the building of a logistics city is an important event in the production of urban space and infrastructure dedicated entirely to supporting networked flows across global supply chains, while on the other hand, the very premise of protecting those flows from disruption entails new forms of political geographic enclosure. For a city dedicated to logistics the interruption of flows becomes system vulnerability and forces that interfere with flow are managed as security threats. According to the logic of supply chain security, threat can just as easily come from a labor action as a natural disaster or a terrorist act. In a moment at once defined by national securitization and global trade intensification, the extreme means of managing this critical dilemma of flow and containment in Dubai Logistics City has become a model incrementally replicated around the world. Accumulation through dispossession of formal and substantive citizenship rights, managed through the production of securitized logistic space, is the ominous model that U.S. port cities are borrowing from Dubai.

This chapter argues that global logistics has recast the meaning and practice of national security, such that preventing the disruption of commodity flows and protecting transportation infrastructure is giving rise to new forms of urban space—the logistics city. Dubai Logistics City may be exceptional in its particular coupling of frenzied economic activity and anemic political rights, yet it is precisely this exceptional form that is serving as the model for the protection of infrastructure and trade flows and reshaping ports in the United States. The chapter will further argue that this model has profound implications for labor and citizenship. The challenge of reconciling global flows and secure nations is giving rise to a rebordering of urban space and the creation of exceptional zones of securitization. On the one hand these spaces intensify the politics of nationality by instituting new practices and hierarchies based on citizenship status, while on the other hand they undermine traditional sovereign state security through the deliberate blurring of military/civilian authority. Perhaps most importantly, in U.S. ports as in Dubai, economic and political rights are themselves recast as security threats.

FROM LOGISTICS CITY TO HOMELAND CITIES

The UAE may seem a world away from the United States, and plans for Dubai Logistics City likely appear a far cry from the organization of space anywhere in the United States, including its ports. However, Dubai's geographic resolution to the problem of containing disruption and supporting flows is a model for port and infrastructure security in U.S. cities. Direct dealings between Dubai and U.S. ports were called to a halt in 2006 following public controversy in the United States about potential "Arab operation" of American shipping terminals. Massive public outcry denounced a plan for Dubai Ports World (DP World), a UAE state-run firm with terminals in dozens of countries, to assume operation of twenty-two U.S. ports. Like the vast majority of terminals in the United States those in question were already foreign owned and operated. But potential ownership by an "Arab state" provoked broad bipartisan Congressional opposition rooted in fear for national security. The sale was approved by the Executive branch of government, and vigorously supported by the president, who even threatened to use his first-ever presidential veto to protect the deal. Nevertheless, the scale of opposition, peppered with regular expressions of explicit racism, eventually led DP World to sell its interests. Serious concerns about labor practices, and human and civil rights in Dubai occasionally infused the debate; however, the vast majority of public discourse simply equated Arabs with terrorists, using stereotyped and racist imagery. With its Orientalist depiction of shady Arabs sneaking into U.S. ports—the *New York Times* cartoon below from February 2006 by an animator who was heavily critical of the Bush administration—captures well the racist tone of popular debate (Figure 5.2).

Nevertheless, the failure of this deal should not blind us to the broader connections between these places and their ports. While direct ownership has been abandoned, political forms still migrate. Dubai has long been touted as a model of success that the United States and the rest of the world should follow. In 2005, former president Bill Clinton said, "Dubai is a role model of what could be achieved despite the other negative developments in the region. Look at Dubai, which has achieved enormous economic growth in such a short period of time." Then president George W. Bush echoed this sentiment on a visit to the UAE. In January 2008 he exclaimed, "I'm most impressed with what I've seen here. The entrepreneurial spirit is strong, and equally importantly, the desire to make sure all aspects of society have hope and encouragement."

Dubai is specifically touted as a model for U.S. port security. To "American officials, the sprawling port along the Persian Gulf here, where steel shipping containers are stacked row after row as far as the eye can see, is a model for the post-9/11 world," Fattah and Lipton assert.[315] "Fences enclose the port's perimeter, which is patrolled by guards. Gamma-ray scanners peek inside containers to make sure they carry the clothing, aluminum, timber and other goods listed on shipping records. Radiation detectors search for any hidden nuclear material," they continue. Leading maritime security experts and logistics trade publications reiterate the ironies of the failed DP World deal in the United States: despite populist concern, Dubai is a global leader in port security.[316]

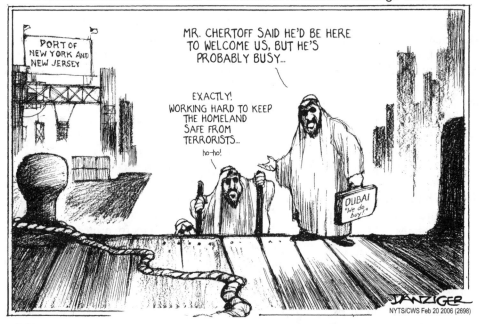

Figure 5.2 Cartoonist, Jeff Danziger's Response to Attempts by Dubai Ports World to Take Over US Ports in 2006, Source: Jeff Danziger, 2006

Dubai Logistics City is an example of how a plan for urban space can both represent and also engineer new political forms. Plans for urban space are crucial in assembling governing visions and practices, and from time to time, a plan emerges which captures something particularly revealing of shifting political rationalities. In a set of recently translated lectures, Foucault declares that he wants to examine "spaces of security" as vital to the constitution of changing political forms, and immediately asserts, "obviously, I will look at the case of towns."[317] For Foucault, master plans for urban space are fragments that allow us to examine changing forms of political rule, from sovereignty to discipline to security. Looking to a series of historical plans for urban space, Foucault asserts that sovereignty literally "capitalizes a territory, raising the major problem of the seat of government, whereas discipline structures a space and addresses the essential problem of a hierarchical and functional distribution of elements, and security will try to plan a milieu in terms of events or series of events or possible elements of series that will have to be regulated within a multivalent and transformable framework."[318] As Elden asserts, "the spatial distribution for sovereignty, discipline, and security is equally important but differently organized."[319]

Today, the plan for Logistics City responds to the pressing contemporary challenge of securing global flows locally. A city dedicated to logistics raises that challenge more directly than any other urban form, and provides a "solution" with broad application. The plan for Logistics City works to design out unpredictability and interference from the smooth space of flows. Certainly the goal of uncontested order and efficiency is an impossible project—a fantasy of state and corporate managers that could never be fully realized. Projects like this are invariably contested and much messier than the sharp lines envisioned for space and social practice. However, while models and diagrams may never be practiced

as planned, they are nevertheless tools for acting upon and organizing life. For Lefebvre, the work of powerful actors in crafting models and representations of space—what he terms "conceived space"—is vital to the production of lived space. In other words, "representations of space are certainly abstract" but they also "play a part in social and political practice."[320] Perhaps most importantly for Lefebvre, "representations of space are shot through with a knowledge (savoir)—i.e. a mixture of understanding and ideology." Huxley also asserts the importance of plans when she argues that we should not dismiss them as "expressions of naive or mistaken spatial or environmental determinisms" because they play "an important part in shaping practices of regulation and management of the urban."[321] Elden further argues that plans and diagrams "serve as models, tests and ongoing aims against which programmes of government are evaluated and adjusted, with the continuous (but seldom attained) aspiration that reality can be made to conform to the truth of these schemes."[322]

So what is specific to the ideal plan for Dubai Logistics City that makes it significant to port security in the United States? On the one hand, there is nothing particularly novel about the project of containing disruption and facilitating flows. As Christine Boyer has argued, since the emergence of the American metropolis in the post Civil War era, two problems defined its government: "how to discipline and regulate the urban masses in order to eradicate the dangers of social unrest, physical degeneration, and congested contagion, which all cities seemed to breed, and how to control and arrange the spatial growth of these gigantic places so that they would support industrial production and the development of a civilization of cities."[323] Indeed, Boyer suggests this problematization of city space and the "quest for disciplinary control" forged a "new relationship between the urban public and social science knowledge, as well as the architectural adornment of urban space and the rational treatment of spatial development." In the process this struggle for the city gave rise to the field of urban planning. Boyer argues that disciplinary order "begins with a fear of darkened places of the city," spaces that should "open to light and ventilated by fresh air." Discipline, "proceeds from the distribution of individuals in space."[324]

If the general problem of reconciling sociospatial order with efficiency and productivity through urban space has a long history, it takes on a more specific form when cast through the geographies of logistics and the politics of "security." Melinda Cooper suggests that Foucault's account of the rise of urbanism is promising because it focuses "on circulation rather than the localization of power, but also because it suggests a genealogy of the event and its relationship to infrastructure."[325] Indeed, across these different time-spaces of government, "the problem of the town was essentially one of circulation."[326] This problem of circulation is central to the whole enterprise of logistics, and the fact that it is conceptualized not as an element of urban form but as an urban form unto itself is indicative of the building of Dubai Logistics City as a significant event. Logistics City is a reflection of the rise of logistics over the past few decades from a residual to a leading concern of business strategy, and a part of institutionalizing logistics at the core of globalized production, trade, and *security*.

DISABLING DISRUPTION

Like Dubai, U.S. ports have also become transnational logistic centers. Ports are intermodal hubs where millions of containers enter the United States every year by sea and are transferred to rail or truck to be distributed to warehouses or production and consumption centers across the country. Rapid movement from primary production sites, largely in China, across multiple modes of transport and through a series of transfer points in places like Dubai, take commodities through U.S. ports and on to American consumers. With so much cargo traveling overseas—90 percent of global trade and 95 percent of U.S. bound cargo—ports are critical nodes for managing supply chains. For the same reason, U.S. ports are subject to intensive securitization. As global commodity flows have become increasingly dependent on speed, they have also become acutely vulnerable to disruption. These flows support a form of accumulation built on just-in-time delivery systems that operate by cutting time and cost. But this precision also produces precariousness. When timing is so tight, minor delays can have ripple effects of paralyzing proportion. As the Organization for Economic Cooperation and Development asserts, world trade is fundamentally dependent on a system of maritime transport that has been made "as frictionless as possible," which renders that system vulnerable, as "any important breakdown in the maritime transport system would fundamentally cripple the world economy."[327] Indeed it is new security measures that are transforming U.S. ports into bounded spaces of exceptional government, inching them toward the Dubai model.

This dilemma puts ports high on the agenda, such that each of eleven plans developed after 9/11 to support "supply chain security" target maritime and port security.[328] Ports are now at the center of national security debates, with the competing demands of "economy" and "security" framing policy design. As Admiral James Loy of the U.S. Coast Guard asserts, "security barriers can easily become trade barriers. To sustain prosperity, we open the gates. To ensure security, we close the gates. We clearly need to get beyond the metaphor of an opened or closed gate."[329] Countless policy documents and reports from military and civilian security agencies echo this conundrum.

In response to these new challenges of

securing both movement and borders, post 9/11 securitization efforts have entailed a dramatic rethinking of the meaning and practice of security. Only nine days after September 11, 2001, the U.S. government began a massive reorganization of domestic and international security under the rubric of the "War on Terror." This took place through the creation of the Office of Homeland Security and the reorganization and expansion of security institutions within the United States, as well as through international action. The U.S. government demanded sweeping regulatory change from foreign states. In some instances, the United States directly defines elements of security practice for other countries and elects itself as enforcement, whereas in other cases, American officials have pressured supranational governing bodies to develop new policies where noncompliance results in isolation from global trade.

The Container Security Initiative (CSI) is an example of the former (Figure 5.3). It posts U.S. customs officials in foreign ports to inspect U.S.-bound cargo, and aims to "extend [the U.S.] zone of security outward so that American borders are the last line of defense, not the first."[330] Dubai was the first "Middle Eastern entity" to join the Container Security Initiative (CSI) in 2005, and as a result, the U.S. Border Protection (CBP) works especially closely with Dubai Customs to screen containers destined for the United States. As the CBP reports, "cooperation with Dubai officials has been outstanding and a model for other operation within CSI ports."[331]

On the other hand, the International Ship and Port Facility Security (ISPS) code was defined and is administered by the United Nation's International Maritime Organization. Crafted at the direct behest of the United States in 2002, it defines basic standards of security with which international ports and ships must comply. In 2004, the code came into effect globally, was adopted by 152 nations, and now requires the compliance of 55,000 ships and 20,000 ports.

These programs are elements of the Department of Homeland Security's (DHS) conception of "layered security." Recognizing

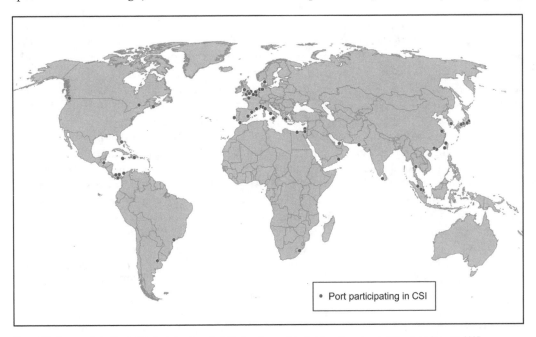

Figure 5.3 Ports participating in U.S. Container Security Initiative. Source: Adapted from Department of Homeland Security, 2007

the geographic complexity of contemporary insecurity, "layered security" acts on a variety of "fronts" to (1) keep dangerous cargo out, (2) address "infiltration" from within, and (3) to secure infrastructure. One single program is at the center of these efforts yet has remained well under the radar of popular debate.[332] Designed to work across all three "layers" of security, the Transport Workers Identification Credential (TWIC) is a key technology in the reordering of urban space, labor, and citizenship. The TWIC presumes that targeting a key link in supply chains—the people handling cargo in ports—will preempt disruption.

Containing labor is a key realm of convergence between Dubai Logistics City and U.S. ports. A violent refusal of workers rights in the interests of trade flows is crucial to the Dubai model, and the TWIC program can be understood as an attempt to institutionalize similar violence. Labor actions in U.S. ports have recently been cast as security threats. The most dramatic instance came in 2002 when George W. Bush enacted the Taft-Hartley Act—America's notorious "slave labor law," in the largest U.S. ports of Los Angeles/ Long Beach. Protest at the increasing numbers of workplace injuries by the International Longshore and Warehouse Union (ILWU) and a subsequent lockout by employers led the U.S. Vice President to declare ILWU actions a *threat to national security*. Following pressure from corporations such as Dell, Ford, and Boeing, who experienced shortages in their just-in-time supply chains, President Bush compelled workers to comply with Pacific Maritime Association (PMA) demands under threat of fines, criminal charges, and even military deployment.

The TWIC aims to preempt such disruption. Designed by DHS's Transport Security Administration, TWIC is managed jointly with the Coast Guard, and operated by Lockheed Martin. The program creates special "security zones" around ports where normal civil and labor law is suspended. Access to "secure areas" is controlled by biometrics and security clearances, organized through extensive new fencing, security gating, cameras, and other surveillance technologies. Secure areas are governed differently; political, social, and economic rights that are in force across the country are suspended within these exceptional zones in the name of national security.

In order to access the port, workers must undergo invasive security screenings, which include criminal history records checks, immigration checks, and intelligence/terrorism checks. If successful, they must carry biometric security cards that link to security perimeters surrounding ports (Figure 5.4). Security clearance and so employment can be easily denied on a permanent basis for a long list of acts including the "attempt to improperly transport a hazardous material" or the "attempt to commit a crime involving a security transportation incident." Clearance is denied for at least seven years for a much longer list of acts including "attempted dishonesty, fraud, or misrepresentation, including identity fraud and money laundering," attempted immigration violations, and "attempted distribution, possession with intent to distribute, or importation of a controlled substance." The TWIC affects at least 1.5 million workers, and there are plans to expand the program across supply chains, well beyond its incubation in the ports.

An estimated 30 to 50 percent of port truckers who are undocumented migrants will automatically be ineligible for the pass. At first glance this would suggest that formal citizenship status will be more important in accessing the ports, but in practice this may not be the case. Responding largely to Wall Street fears of a shutdown of the ports without this crucial supply of low cost undocumented labor, senior DHS officials insist that alternative solutions to the dilemma will be developed. Ports might extend "assumed identities" to undocumented truckers and turn an organized blind eye to their missing

Figure 5.4 Transport workers identity credential: A biometric identity card.

status, or even establish container pickup areas just beyond secure areas. This suggests that the program is unlikely to remove undocumented workers from the supply chain, but will perhaps make all port labor more precarious.

The TWIC rewrites both the limits of state surveillance and supplants labor protections, but does so without presenting itself as labor law. It undermines collective agreements and privacy rights, and evacuates any meaningful notion of employment security. Most importantly, the TWIC blurs the border between crime and terror, and between police and military authority. These borders between different authorities and legal codes for governing insecurity are fundamentally geographic; they are the once sacred borders of national sovereignty. New security programs are both a response to changing conceptions of security and a key part of the transformation of legal and social technologies for governing border spaces. This recasting of ports outside normal national space is part of a broader reconceptualization of the geographies of security. In a moment where national security is no longer conceptualized as independent of global trade, urban port space has become

a key target for reform. This complex set of spatial and scalar relations suggests that there are some vital relationships between supply chain security, scale, and cities that this chapter will turn to explore below.

Reworking Security, Net-Working Security

> The term "port security" serves as shorthand for the broad effort to secure the entire maritime supply chain, from the factory gate in a foreign country to the final destination of the product in the United States.[333]

The above quote hints at some dramatic shifts that are underway in the geographies of emerging conceptions of "security." Since the end of the Cold War, both military and civilian agencies have been actively rethinking security in order to respond to the changing nature of threat. For the geopolitical state this means the division of inside/outside ordered authority, jurisdiction, and rights, both geographically and ontologically, with the foundational exception of colonialism. The border as territorial line was the basis for the division between police/military, war/peace, and crime/terror. Formal citizenship rights were also created anew through nationalization and we have come to naturalize the relationship between citizenship and nationality as a result. However, concern for the security of supranational systems trespasses these sociospatial forms, as the distinction between "internal" and "external" security becomes more difficult to discern. This is starkly evident in the increasing entanglement of police and military jurisdiction and of crime and terror in the ports.[334]

The challenge to the geopolitical border reorganizes space and scale. Perhaps most importantly, it puts cities in a different relationship to security. In this "networked" notion of national security the territorial border can be a problem rather than a solution, and cities become crucial nodes for managing flows. For Collier and Lakoff, the pressing questions for critical scholarship are "what type of security is being discussed? And what are its political implications?" rather than the common debate regarding whether we have "too much or too little."[335] They present a genealogy of collective security that traces the emergence of different forms, each "characterized by a distinctive way of identifying threats and organizing mitigation measures."[336] Drawing on Foucault's work they suggest that "sovereign state security" dates back to modern nation-state building, responds to concerns for territorial and juridico-legal integrity, and mobilizes military force or diplomacy in response to threats. Sovereign state security was joined (not replaced) by another form of security in the late nineteenth century, most powerfully evidenced in the Keynesian welfare state. Population aims to secure the health and welfare of the social body, and mobilizes techniques like social insurance and social welfare.[337] They introduce a third, more recent form of collective security, which I suggest corresponds with Foucault's discussion of the rise of "security" in his investigation of urban space. "Vital systems security" emerged in the first half of the twentieth century in the context of mass war. It seeks to protect systems that are critical to economic and political order ranging from transportation to communications, water supply, and finance, against threats that may be impossible to prevent "such as natural disasters, disease epidemics, environmental crises, or terrorist attacks." Vital systems security is distinguished by the wide range of disasters to which it aims to respond, and by its emphasis on preparedness for emergency management rather than predictive responses that characterized risk-based models of insecurity. "Vital systems" is also distinguished by a geography that is explicitly networked, as opposed to territorial in form. "Vital systems" are the networks and infrastructures that function by virtue of their connectivity, and this connectivity is typically supranational.

Kim Rygiel has made parallel arguments regarding the "securitization" of citizenship through new border control and detention arrangements. "While such policies are about securitizing the social body on which traditional notions of state security depend, the implementation of such measures has not led simply to the re-entrenchment of the modern territorial state," she argues.[338] "Paradoxically, it is precisely this process of securitizing citizenship (through various forms of border controls and detention practices) that has facilitated the development of citizenship into becoming a globalizing regime of government." Rygiel suggests that through the work of national security policy, citizenship has become "as much about governing populations between and across states as it is about governing within the state."

Indeed, military and civilian authorities are proposing these kinds of transnational conceptions of security that rely on new models of border space. Common across many military and civilian designs is not the suspension or dismantling of the border, but the creation of a different form and geography of border management. The U.S. "smart border" aims precisely to facilitate flows of certain kinds, while "interdicting and stopping 'bad' people and 'bad' things from entering the country."[339] In this new paradigm, the border is transformed from a two-dimensional line across absolute space, to a transitional zone, defined by exceptional forms of government that overwhelm established jurisdictions and spaces. State security experts insist that old categories are creating problems for law enforcement and international security work, and it is precisely the blurring of established categories, jurisdictions, and spaces that must guide responses to insecurity today. Recent work by RAND targets the border for reform and focuses specifically on the maritime border. In order to support their role as nodes in global systems, RAND argues that ports must be governed as spaces in-between national territories.[340] In 2006, U.S. Army

Lieutenant Colonel, Thomas Goss,[341] called this new border space "the seam." The seam figures as a zone between inside and outside national space, where old geopolitical divisions no longer hold; the border between police and military authority is blurred, and so too is the line between crime and terror. Like the OECD and RAND, Goss uses the maritime border as the paradigmatic space for experimentation and reform precisely because of the magnitude of the challenge of "opening and closing" access to trade flows.

While this experimentation with border space may undermine simple territorial notions of security, it also rejuvenates security through its transformation. The conception of the border as a "seam"—a special zone rather than a bifurcating line, aims to rework the meaning and practice of security to support systems that often span national spaces. And indeed, the TWIC program aligns quite perfectly with this vision. TWIC does the work of bounding ports and at once transforming them into distinct spaces of government, while blurring the lines of inside and outside and of crime and terror, within: TWIC transforms U.S. ports into exceptional seam spaces.

Supply Chain Security and Logistic Space

Network or "systems" notions of security like those described above are highly relevant to contemporary securitization initiatives and offer rich insight to an analysis of the ports in particular. The diagram below, with its multiple and connected sites of activity and insecurity, illustrates how a network approach to security fits the challenge of global supply chains. However, it is not just any system, but *economic systems* that are driving some of the most urgent change in models and practices of security. In the current moment of neoliberal globalization, supply chains assume a particularly pivotal role in the recasting of security. Countless scholars have argued that the ascent of

neoliberal policies, political economies, and governmentalities has seen economic calculation engulf other domains. As Nikolas Rose argues, neoliberalism demands that "social government must be restructured in the name of an economic logic…all aspects of social behavior are now reconceptualized along economic lines."[342]

Across a wide variety of government action—particularly in the area of urban development and infrastructure—economic measures of value have become political common sense as business rationale has reshaped "public interest."[343] Indeed, national security is now frequently conceptualized as almost interchangeable with the security of supply chains, with national interest and global trade wed seamlessly in experts' minds, if not in practice. Not only is a "strong world economy" supposed to enhance "our national security by advancing prosperity and freedom in the rest of the world" as George W. Bush outlined in the National Security Strategy of the United States of America. In the world of border management, the security of global supply chains is vital to national security, with the former often even conflated with the latter. Supply chains play a vital role in the rise of networked notions of security and so an investigation of its emergence and effects demands a genealogy of the field of logistics (Figure 5.5).

But what *is* logistics? Why is it so relevant to questions of disruption and in/security and how does it redefine urban space? Logistics is now ubiquitous in global trade such that the genealogies of its practice and the politics of its spatial calculation are hidden in plain view. The word *logistics* appears on virtually every big rig truck on almost every major highway in the world, and is used innocently in its vernacular sense to simply mean the nuts and bolts of "getting something done." But logistics has a more precise past and present than all this suggests.

A science of the efficient design and management of supply chains, foundational to

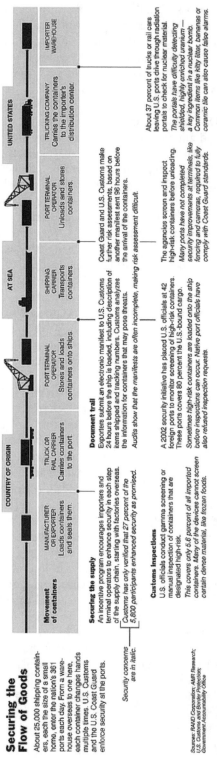

Securing the Flow of Goods

About 25,000 shipping containers, each the size of a small home, enter the nation's 361 ports each day. From a warehouse overseas to one here, each container changes hands multiple times. U.S. Customs and the U.S. Coast Guard enforce security at the ports.

Security concerns are in italic.

Securing the supply

An incentive program encourages importers and terminal operators to enhance security in each step of the supply chain, starting with factories overseas.

Customs has only verified that 27 percent of the 5,800 participants enhanced security as promised.

Customs inspections

U.S. officials conduct gamma screening or manual inspection of containers that are designated high-risk.

This covers only 5.6 percent of all imported containers. Many of the devices cannot screen certain dense material, like frozen foods.

Sources: RAND Corporation; AMR Research; U.S. Customs and Border Protection; Government Accountability Office

COUNTRY OF ORIGIN

Movement of containers

MANUFACTURER OR EXPORTER
Loads containers and seals them.

TRUCK OR RAIL CARRIER
Carries containers to the port.

PORT TERMINAL OPERATOR
Stores and loads containers onto ships.

Document trail

Exporters submit an electronic manifest to U.S. Customs 24 hours before the ship is loaded, including description of items shipped and tracking numbers. Customs analyzes the information for containers that may pose threats.

Audits show that the manifests are often incomplete, making risk assessment difficult.

A 2002 security initiative has placed U.S. officials at 42 foreign ports to monitor screening of high-risk containers. These ports covers 80 percent of the U.S.-bound cargo.

Sometimes high-risk containers are loaded onto the ship before inspections can occur. Native port officials have also refused inspection requests.

AT SEA

SHIPPING CARRIER
Transports containers.

Coast Guard and U.S. Customs make further risk assessments, based on another manifest sent 96 hours before the arrival of the containers.

The agencies screen and inspect high-risk containers before unloading.

Many ports have not completed security improvements at terminals, like fencing and cameras, required to fully comply with Coast Guard standards.

UNITED STATES

PORT TERMINAL OPERATOR
Unloads and stores containers.

TRUCKING COMPANY
Carries the containers to the importer's distribution center.

IMPORTER WAREHOUSE

About 37 percent of trucks or rail cars leaving U.S. ports drive through radiation portals to check for nuclear material.

The portals have difficulty detecting shielded, highly enriched uranium — a key ingredient in a nuclear bomb. Common items like kitty litter, bananas or ceramic tile can also cause false alarms.

Figure 5.5 Securing the supply chain: Source: New York Times, 2006.

the very possibility of just-in-time production systems and the global economy they have helped to build, logistics is devoted to efficiency and speed. But logistics was not always what it is today. Historically logistics was a military art rather than a corporate science—one of the three arts of war along with strategy and tactics in the Napoleonic era.[344] Logistics did not become subject to sustained civilian business interest until the post-World War II period.[345] This was in part due to the tremendous advances in military logistics that occurred during the war, as well as the growing complexity and changing nature of production, distribution, and consumption in a rising "consumer society." A flurry of applied writings on the topics of "distribution," "rhochrematics," and eventually "business logistics" circulated through the 1960s, provoked by a growing concern for cost recovery in business operations in connection with the recession of the 1950s.[346] Key works in this period emphasize the rationalization and deliberate management of spatial organization within the firm, as well as the opening up of a new space of action—the geographic and calculative space of operations between the end of the production line and the point of commodity consumption.

By the early 1960s, cost minimization had been replaced with a model that emphasized value-added.[347] The shift to a profit maximization approach, which followed the introduction of systems theory into the field of distribution geography, transformed logistics. Decisions were no longer to be made based on reducing costs at the scale of one action between two points, but rather in terms of raising profit across the entire production "system" up to the point of commodity consumption. From this point onwards, logistics became a *science of systems* and its more circumscribed concern with distribution transformed into an umbrella science of spatial management.[348] It was this "systematization" of distribution that gave way to "the revolution in logistics," which was a key underpin-

ning for the globalization of supply chains. This revolution in logistics was a momentous event in the discursive and material reorganization of economic space at the global scale which today demands new forms of border control and meanings of "national" security.[349]

Today, logistics is a powerful growth industry. Vertical integration has been rapid with firms that once specialized in one component of production or distribution assuming new corporate identities in logistics, acting as full service systems managers of global supply chains. Through the 1990s, new professional associations formed and enrollment expanded in a growing number of university and college programs. Trade magazines that once catered to shipping, distribution, or materials management now orient themselves to logistics, while supply chain management, a core element of business logistics, became a growing field in business and management schools, and has even begun displacing traditional economics departments.[350]

The revolution in logistics also entailed dramatic restructuring of *urban* space. The world's largest corporation, Wal-Mart, is a leader in logistics calculation and has remade marketing strategy and urban space. The rescaling of logistics from a national to a global strategy has produced the rise of big box retailing and so the accelerated and standardized expansion of landscapes of urban sprawl.[351] But logistic science has recast the urban space of production and distribution, not just consumption. Warehousing districts and intermodal logistics centers exploit labor made captive by deindustrialization and precarious migration and produce spaces that facilitate that control. Logistics calculation has thus played an extraordinarily powerful role in reshaping space and scale; however, the current developments are of an entirely new order. As logistics reworks the space of circulation at global and local scales, the building of Dubai Logistics City reworks the political geographies of security.

LOGISTICS CITIES, AFTER DEMOCRACY

Politics can often be profoundly antipolitical in its effects: suppressing potential spaces of contestation; placing limits on the possibilities for debate and confrontation. Indeed, one might say that one of the core functions of politics has been, and should be, to place limits on the political.[352]

At a time when the economy is said to be vital to national security, and when national economies have become so thoroughly entwined in global trade, it is perhaps not surprising that the protection of global supply chains has become vital to national security strategy. Disruption to commodity flows exposes the vulnerability, even precariousness of just-in-time production systems and so too the centrality of the transportation infrastructure and its protection to the politics of our present. If the migration of transportation infrastructure to center stage in the contemporary politics of security is now common sense, then perhaps more surprising is the way in which the city and citizenship are targeted in efforts to preempt their disruption. The Logistics City exposes defining tensions and transformations in contemporary political geography; the city becomes a space for managing national security within global supply chains. As the maritime border is reconceptualized as a "seam" in supranational networks, the resulting dilemmas of containment and flow are foisted upon port cities.

When disruption of flow is cast as a problem of security, then a motley crew of disruptive forces may be brought under the purview of securitization. Indeed, natural disasters, terrorist acts, and labor actions are all now governed as threats to supply chain security. But it is the conflation of the latter two on this list that is the most troubling, and which raises profound questions about the future of politics and democracy in a context where economy and security are so tightly entwined. That labor actions in U.S. ports are now regularly cast as threats to national security, even "terrorist" acts, is one clear sign of how deeply entangled economy and security have become. It also reveals how powerfully this entanglement can undermine claims to democratic freedoms, labor, and citizenship rights.

Dubai Logistics City offers the world an *antipolitical model of economy*,[353] and so too of space. Logistics City does not define itself as a political invention or innovation, but is clearly deeply political in the radical ways it recasts relationships between space, status, security, and labor. Rather than being a stated goal of Logistics City, shrinking the space of political claims is an explicitly unspoken effect; politics becomes "overflow" in this field.[354] Urban space built entirely for the purpose of supporting supply chains, the "antipolitical economy" of logistic cities takes economic calculation to new heights. Dubai Logistics City frames itself entirely in business language; "customer service" stands in for the relationships between the city and its "citizens," with the exception of workers who are governed as disposable potential security threats. And while it is neoliberalism that is typically charged with "the explicit imposition of a particular form of market rationality on [political] spheres,"[355] logistics foregrounds the geographies of these market rationalities. Logistics transforms spatial calculation into cost-benefit equation while supply chain security works to preempt disruption. Together they are remaking security, citizenship, and space.

And yet, while Logistics City aims to see the economic literally *displace* the political, city plans that present political projects through physical design are never simply transposed into practice.

At stake in the Logistics City is the politics of securing the infrastructure of economic space. The securitization of U.S. ports is crystallizing logistic logics in "the Homeland." Programs like the TWIC govern ports as exceptional seam spaces of economic flow that are literally outside the space of normal national law; they work politically to depoliticize the space of economic flow. U.S. ports are transformed into logistic cities; however, this transformation is far from complete. While Logistics City may well be a model for the state and capitalist actors that govern logistic space, efforts to actualize it are highly contested, particularly by workers whose lives are so directly constrained by this architecture of flow. In Dubai, workers take action even at the risk of permanent expulsion, and in the United States, longshore workers risk the label of "terrorist" to contest this future.

This contestation does not simply oppose or resist logistic spatial logics, however, but transforms logistics from a technical science into a field of political struggle. For instance, the west coast International Longshore and Warehouse Union has recently initiated its Global Logistics Institute, a progressive think tank geared precisely to generating worker-centered logistic knowledge. The Logistics City certainly raises enormous challenges for democracy and the movements of workers that claim democratic rights in seam spaces, but it has also become a new field of politics and struggle, not the site of their death.

CHAPTER 6

Clogged Cities: Sclerotic Infrastructure

Simon Marvin and Will Medd

While strategies to tackle obesity have led to renewed debate about the specific relationship between the body and urban forms, the deposition of fat through bodies, cities, and infrastructure reveals a more complex web of urban metabolisms.[356] We argue that in order to understand the circulation of fat in a city context, metaphors of urban metabolism become important. Urban metabolism need not refer to stable sets of relationships to which explanations of social order subsequently refer: consider, for example, the early Chicago School work.[357] However, to reject the concept of metabolism by reducing it to functionalist and teleological metaphors would be to lose the insights a reformulated concept can reveal. "Cities," David Harvey argues, "are constituted out of the flows of energy, water, food, commodities, money, people and all the other necessities that sustain life."[358] Metaphors of metabolism are therefore useful for understanding such flows. The contingencies and mobilities of fat in bodies (as individuals), cities (as a collective site of action), and sewers (as infrastructure), we argue, highlight a multiplicity of urban metabolisms, each with different interconnectivities and forms of instability.

Fat poses key challenges for understanding the future of contemporary society. For, although fat can appear very much as a mobile fluid, traveling through global networks of food production and consumption, it also encapsulates the other aspect of contemporary networked cities: immobility. Fat literally, but also metaphorically, can also appear as static, difficult to move, and solidified. Although fat provides bodies with important functions (for example, cushioning, insulation, and storage) it is the excess of fat that causes so much concern.[359] More specifically excess fat, particularly when bodies become obese, is associated with an impressive list of associated health problems.[360] At a population level, internationally there is much concern about the growing levels of obesity. For example, the International Obesity Task Force (IOTF) reports data illustrating obesity concerns in a wide range of countries including Australia, Japan, Brazil, Britain, the United States, Mauritius, Kuwait, and Western Samoa. The IOTF was established to "alert the world about the urgency of the problem of the growing health crisis threatened by soaring levels of obesity," signifying the growing international anxiety about the global epidemic of obesity.[361]

In this chapter we examine how, in response to the rising numbers of obese bodies, there has been the mobilization of the concept of fat cities involving renewed debate about the relationship between bodies and the city, provoked largely by the innovative representations of a men's fitness magazine. We shift focus in this debate to look to the problems of fat in infrastructure, focusing specifically on the experience in U.S. cities of sewer blockages that reveal quite different sets of processes within which fat is embedded. We show how in each of these sites of intervention the body, the city collective, and the sewer strategies of prevention, removal, or reuse each reveal a multiplicity of metabolisms as well as partial interconnections between them.

THE EMERGENCE OF OBE*CITY*

The proliferation of anxiety about the obesity epidemic is in part mobilized by the experience in the United States. It is reported that 60 percent of the U.S. population is overweight and 21 percent obese.[362] This percentage has more than doubled in the last two decades, with obesity now rising by 5 percent per annum. This is not evenly spread through the population. There are serious health inequalities with more than half of black women in low socioeconomic groups being obese. One in five children is overweight. Obesity is estimated to account for 12 percent of health care costs, $100 billion a year and rising. The United States has provided the emblematic marker for the rest of the Western world to signify where other countries are heading and as a context to examine the causes of obesity, the problems it generates, and possible solutions. In this chapter, however, we take a different turn to examine the crisis of fat deposition that has started to raise the visibility of the interconnections between the multiple metabolisms of the body, sewer, and city. In the United States, the concern with the rising levels of obesity of individual bodies has sparked an interesting mobilization of the problem of obesity at the urban level.

At the core of this movement has been the development of city fatness and fitness league tables produced by *Men's Fitness*, a leading men's health magazine. The magazine set out "to measure, city by city the relative environmental factors that either support an active, fit lifestyle, or nudge people towards a pudgier sedentary existence."[363] The magazine's analysis enrolled a range of existing surveys and data and involved aggregating scores based on a multiplicity of indicators, including gyms and sporting goods stores, health club memberships, surveys of exercise, levels of fruit and vegetable consumption, alcohol consumption, smoking, television watching, number of junk food outlets per 100,000 population, and recreation facilities. Obesity levels were scored for cities by drawing on data from the Centers for Disease Control and Prevention and the Center for Chronic Disease Prevention and Health Promotion. Consequently cities have been jostling not to be top of the league. Whereas New Orleans was ranked as the fattest city in the first year of the league in 1999, in 2000 it was Philadelphia, and then Houston took the lead. These rankings were not ignored.

An ABC news headline announced "Houston you have a problem: a big, fat problem" and for 2002 and 2003 Houston was at the top of the Fat City league table.[364] This identity was provocative, not something Mayor Lee Brown wanted "as a distinction of our city." Houston's response was to get advice from Philadelphia. A Fitness Czar was appointed, Lee Labrada, a former Mr. Universe who began a "Get Lean Houston" campaign. This included a "Fat Drive" in which participants pledged to lose weight, resulting in 17,000 pounds being lost by 2,000 flows of fat through bodies, cities, and sewers, with 315 participants by 2003. Again a range of different agencies was enrolled. Most notable was MacDonald's as Official Restaurant Sponsor rolling out a menu of "Salad and More" across all 253 of its restaurants in Houston. By 2004 Houston was able to boast "Houston: We did it! We are officially no longer the Fattest City in America. Dropping to #2 behind Detroit, MI."[365] This, according to *Men's Fitness,* was the result of better scores for increased sports participation, decreased alcohol consumption, and improved nutrition.[366]

The mobilization by *Men's Fitness* of the

metaphor *fat city* has provoked a renewed interest in the relationship between bodies and urban form.[367] Papers published in *The American Journal of Health Promotion*[368] and *American Journal of Public Health*[369] reported a series of studies that are among the first to link shopping centers, lack of sidewalks and bike trails, and other features of urban sprawl to "deadly" health problems. One report shows how "people who live in sprawling neighborhoods walked less and had less chance to stay fit...people living in sprawling neighborhoods weighed 6 pounds more on average than the folks living in compact neighborhoods where sidewalks are plentiful and stores and shops are close to residential areas."[370] And, in such areas, obesity is reported as more prevalent. Daniel Sui argues this literal uptake of the fat metaphor can be illuminated through debates about the body and the city:

> Fat City depends on the fatness of the body since this is the body of a coordinator: a data processor, business person, real estate agent, or urban planner.... And the fatness of the body depends on the fatness of the city, since it develops as a result of the automobile dependent, privatized spaces of the fat city. The excess circulation of the city (roads) allows isolation of the person, which transforms again into excess circulation (blood vessels to the fat tissue). If the fat body becomes an obsession for health reasons and narcissism alike, it cannot be returned to a more sustainable size and shape as long as the city remains outside of theories of health.[371]

Raising analysis of the relationship between the body and the city in this way is important; however, the emphasis on the direct linkages between fat bodies and urban sprawl neglects the subtleties and multiplicities in the work done by *Men's Fitness* magazine. The *Men's Fitness* league tables have significantly raised the visibility of fat issues in U.S. cities, where it has mobilized the classic debate between nature versus nurture by arguing that the rapid rise in obesity in the last century could not simply be explained by genetics and must therefore be a consequence of changing environmental circumstances. In explaining its rationale for this it uses the imaginary example of two identical twins living in different cities to argue that the twin living in a poor city environment in which finding healthy food is difficult, where there are few places to exercise, the weather is not conducive to exercise, where people smoke freely, commuting is a hassle, there is poor access to gyms, health care is poor, and where junk food is to hand, is more likely to be fat. With this simple analogy to demonstrate the issue, the magazine develops a sophisticated methodology upon which the league tables are based. The magazine displaces methodology based on simple measures of obesity prevalence based on body mass index (BMI) with a range of indicators that move beyond individual bodies to the wider city ecology. This is clear when one considers that, based on BMI alone, the league tables would look different. For example, in 2003, Memphis, Tennessee, had the highest percentage of overweight and obese adults and yet was placed twenty-first in the league tables. The effect is to move from measuring the state of obesity in a city to highlighting the processes that lead to obesity.

This identification of the fat city, as we have seen with the example of Houston, can lead to the mobilization of representations of the city to remobilize fat collectively with the enrolment of a range of diverse actors into strategies to reshape the metabolism of bodies and the city. The emergence of the representation of fat cities has involved a significant move in making visible the dynamics between bodies and the city as an environment and in doing so has opened up new discourses and forms of activity around the city as a collective metabolism. Although *Men's Fitness* magazine presents a sophisticated account of the relationship between fat bodies and the fat city, understanding the circulation and deposition of fat requires looking beyond the dichotomy of body and city. To look at the multiple mediations of fat embedded within other less visible urban metabolisms we need to recognize the complexities and multiplicities of the relationship between bodies and cities that includes the role of social-technical infrastructure in the distribution of fat.

FAT THROUGH INFRASTRUCTURE: THE SEWER-FAT CRISIS

The movement of discourses about fat from the individual to the city level has, as we have seen, involved the mobilization of a range of actors into programs of slimming down American individuals and slimming down American cities. It would be a mistake, however, if the relationship between individuals and cities, as both environment and site for collective action, ignored the role of infrastructures. By its very nature, infrastructure is often defined by its hidden presence with apparent reliability and stability, as if it represents an almost neutral intermediary between the individual and the city. Yet, understand-

ing the city as a sociotechnical process points to the constant effort required to keep infrastructure working.[372] The social, technical, and spatial characteristics of infrastructures become more acutely revealed during times of crisis when the material embodiment of different sets of social, political, economic, or organizational relations are ruptured.[373] This is true of fat and its relationship to sewerage infrastructure: the sewer-fat crisis is a crisis that points to the otherwise largely hidden complex interconnectivities between, for example, the food industry, local waste-disposal systems, and global food-oil markets. Sewers, as Matthew Gandy argues "are one of the most intricate and multi-layered symbols and structures under-lying the modern metropolis, and form a poignant point of reference for the complex labyrinth of connections that bring urban space into a coherent whole."[374] A crisis in the sewers is more than a technical malfunction. It is a crisis that reveals the instabilities between the geographies of city infrastructure and the unbounded interdependencies of city metabolisms.

In October 2002, Randy Southerland wrote about the growing numbers of sewer blockages and overflows across cities in the

Figure 6.1 Fat blocking sewer. Source: BBC.

United States as restaurants and fast food chains pour cooking residue into drains and local governments lack the resources to monitor grease disposal and enforce the relevant regulations. The solidification of fat, oils, and grease, he writes, "choke[s] pipelines, eventually clogging them and causing them to cough up rivers of raw sewerage"[375] (see Figure 6.1). He cites how, in January 2001, the U.S. Environmental Protection Agency sued Los Angeles for 2,000 sewer spills over five years, 40 percent of which were caused by fat. In *The Wall Street Journal*, Barry Newman writes how in New York there are about 5,000 "fat-based backups a year with several big gum ups."[376] For reasons that range from the decline in global markets for waste fat and the increased costs of fat disposal, following former Mayor Giuliani's crackdown on the garbage Mafia, more grease is illegally disposed of into the sewers.

> Fat won't pollute: it won't corrode or explode. It accretes. Sewer rats love sewer fat; high protein builds their sex drive. Solid sticks in fat. Slowly, pipes occlude. Sewage backs up into basements or worse, the fat hardens, a chunk breaks off and rides down the pipe until it jams in the machinery of an underground flood-gate. That to use a more digestible metaphor causes a municipal heart attack.[377]

Although there are hotspots around restaurant areas there are also sewer-fat problems around residential areas, particularly where large numbers of multifamily units are located and where residents discharge their grease into the drain. And in some cases attention is turning to schools and prisons.[378] Just as with obesity, then, there are different distributions of sewer-fat problems across different publics (see Figure 6.2).

In contrast to *Men's Fitness*, which made visible the problem at the level of the city collective, making visible the problem in sewers continues to be a challenge for local authorities and utilities charged with sewer maintenance. A variety of techniques have been developed for identifying blockages, including closed-circuit television, smoke infrared thermography, and even radar and sonics.[379] Having made fat visible, cities are developing strategies to combat the fat in the sewers. High-pressure hoses can remove blockages but dislodged blocks of fat may then cause new problems downstream. Large vacuum trucks are also used to either suck or blow fat out of congested sewers. In New York an enzyme product that reduces grease build-up is routinely used, and a liquid emulsifier can now tackle larger build-ups.[380] Similarly in New Bedford, Connecticut bacteria developed by a biotechnological company are used to metabolize the grease, breaking it apart into water, carbon dioxide, and free fatty acids, that can be washed away from metal, concrete, and brick.[381]

When the crisis hits, the problems become all too visible in sewer overflows but prevention is more problematic. Ultimately the concern of cities is to avoid the grease being put into the sewers and making visible the practices of restaurants and households distributed throughout cities raises different challenges. New lean sewer ordinances have been developed, but the cost of monitoring and enforcement is high.[382] Not surprisingly, we see also the emergence of appeals to the city collective once more to reshape the deposition of fat. In New York, for example, the Environmental Protection Department guidance makes the following plea: "Sewer back up damages property and damages public health the city needs businesses and individuals to do their part to maintain the system."[383] Indeed, across U.S. cities brochures promoting "fat-free sewers" show homeowners what they can do to keep grease out of the sewers (see Figure 6.3).[384]

Fat is not just a problem for bodies and the cities' public health, it is also a problem that involves the processes of city infrastructure. The sewer crisis has given us a pertinent example of this, but the arguments can also be understood in relation to other

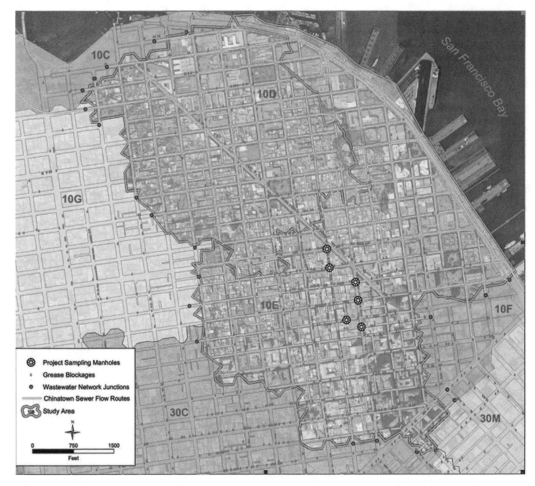

Figure 6.2 Chinatown and Fisherman's Wharf Area Pilot Fats Oil Grease Program, San Francisco.

infrastructures. Indeed, resolving the problem of fat being disposed in the sewers can lead to problems in other infrastructures. A traffic jam in San Francisco, one of the fittest U.S. cities according to the *Men's Fitness* league table, was caused by a spillage of cooking oil from an overturned truck.[385] It took cleanup crews nine hours to reopen the lanes of the highway according to the California Highway Patrol. It was reported that the truck was carrying used cooking oil collected from restaurants. Just as fat in the sewers solidifies as it cools, so too on the highway, in the cool weather, the fat had solidified into a gel making the clean up harder: "it got into the grooves (of the road) and was difficult to extract" a spokesman from the truck company said.

A critical issue then was what to do with the waste fats and oils collected from restaurants and fast food outlets? The EPA has estimated that there are 3 billion gallons of waste oils and fats produced every year in the United States and cities are clearly major producers of such waste.[386] Over the last decade a series of experiments and initiatives have been undertaken, particularly in California, to reuse the waste fats as a fuel. As increased regulation and intervention by local governments and utilities has led to the

Figure 6.3 Fat free sewers. Source: Southern Monterey Bay Discharge Group.

interception of fats the quality of the waste could be increased through monitoring and licensing the waste collection industry. Initially individuals and eco-groups started collecting this waste fat and after undertaking relatively simple conversion methods blended the waste with their diesel fuel. Later the U.S. EPA started supporting community-based biodiesel schemes. For example, in 2005, the U.S. EPA awarded a grant to Ecology Action to pilot the first community-based biodiesel production initiative in the United States:

> This community effort will result in almost 47,000 gallons of waste cooking oil being used to make 190,000 gallons of the B20 biodiesel blend—enough fuel to fill the tanks of over 4,000 city of Santa Cruz recycling trucks, or enough to fuel a fleet of school buses for an entire school district for a year. This project is a model for other cities and counties across the country.[387]

The U.S. EPA is interested in developing fuels from waste oil and fat for three particular reasons.[388,389] First, biodiesel is an alternative fuel that is renewable and can be made domestically. Biodiesel may be blended with conventional diesel to get different blends such as B2 (2 percent biodiesel and 98 percent conventional diesel) or B20 (20 percent biodiesel) or it can be used as 100 percent biodiesel (B100). Although biodiesel costs more than petroleum diesel in some places, the price gap continues to narrow. The price falls considerably when waste cooking oil is used, since 75 percent of the price of biodiesel comes from the oil feedstock. Second, when compared to conventional diesel, biodiesel significantly reduces air pollution emissions of sulphur dioxide, particulate matter (soot), and carbon dioxide. Furthermore, when used cooking oil is recycled to produce biodiesel, billions of gallons of waste grease can be diverted from landfills and municipal water pipes, improving the quality of both air and water.[390]

Increasingly, large U.S. cities can see strategic opportunities in developing fuel from fats in a systemic way that links to their wider environmental and economic objectives (see Figure 6.4). For instance San Francisco has already exceeded its goal of converting 25 percent of the city's diesel fleet vehicles to a biodiesel blend and now has 39 percent of its fleet vehicles running on biodiesel. Norcal Recycling/Waste Management, the city's waste hauler, has converted 100 percent of their fleet to biodiesel.[391] Many other U.S. cities are now looking at the strategic potential of using the metropolitan areas' waste fats and oil as they seek to develop circular or "closed-loop" systems that convert wastes to fuels.

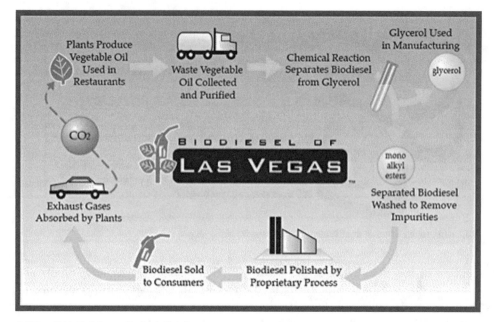

Figure 6.4 From Fat to Fuel. Source: Biodiesel of Las Vegas.

CITIES AS SITES FOR PARTIALLY CONNECTED METABOLISMS

The United States is currently seen as the world leader in the obesity epidemic, as the Land of the Fat.[392] Much attention has turned to the anxiety raised by the U.S. epidemic and the risk that the rest of the Western world is heading in the same direction. Although such attention is usually focused on the problems of fat bodies and associated issues of food and diet, in this paper we have looked to the United States to explore the issues of fat within a specifically urban context. Key to this shift is an understanding of the connections between the problem of fat in individual bodies and the mobilization of fat at the level of the city. Additionally, the embodiment of fat extends beyond fat bodies and fast food into infrastructure itself through the deposition of fat in the sewage network. What then can we learn looking across these contexts?

The city is a site of multiple metabolisms, but these are always partial and selectively connected metabolisms. In this chapter we have not attempted to provide a complete explanation of the political ecology of the metabolism of fat. Consequently, what is missing from our account is the global system of oil and fat production; the international movement and distribution of fats and oils; the production and distribution of food chains that incorporate fats and oils; and the social organization of waste oil and fat systems. At the same time we have not presented a synthesis of the metabolism of fat in terms of the interconnectedness of its social, economic, spatial, biochemical, and cultural dimensions. Yet by using the frame of the city we can do something else: that is, to begin to bring into focus elements of these multiple metabolisms and the partial interconnections between them. What is particularly interesting for us is how strategies for dealing with fat at an urban level start to problematize the metabolisms of fat in quite different ways. There are three strategies for dealing with fat: the removal of fat, prevention of fat deposition, and the reuse of fat as a fuel. These three strategies provide competing ways of seeing the metabolism of fat. Imagine looking through a kaleidoscope that can highlight how each of the three strategies brings into focus the disconnections and partial interconnections between the multiple metabolisms of the city, body, and infrastructure. What we see is a series of changing yet always partial metabolisms and forms of connection.

Removal of Deposited Fat

The first strategy for dealing with the crisis of deposited fat is a strong discourse of removal. These discourses begin disdainfully with unhealthy and corpulent bodies in terms of a representation of excess and greed, are mobilized in relation to the Fat City and Fat Sewers that are to be "slimmed down" or "made lean." And with this discourse the strategies of intervention focus on a range of technological devices used to remove fat from bodies and sewers. These range from suction devices, the physical removal of fat, or the use of drugs and biotechnologies to remove fat. Specialist medical and sewer treatment technologies are mobilized to focus on highly individualized and site-specific fat removal at the level of the individual body and sections of the sewer. This strategy relies mainly on the use of specialist technical and professional expertise with relatively passive users with little collective action at the level of the city. There is a relatively limited understanding of the metabolism of fat within this discourse. Primarily, the focus is on the crisis of blockage and deposition, with the problem being one of extraction and disposal. There is less focus on the wider issues involved in the production, distribution, and consumption of fats and the wider processes that shape its deposition in bodies and sewers. In this way the body and the sewer are decontextualized from their relationships with a wider metabolism of fat. At the same time the interconnections between the me-

tabolisms of body and sewers are relatively weak. The problem of metabolism is deposition and disposal of waste fat.

Within all these contexts the removal of fat is difficult and its disposal expensive. Though strategies for the removal of fat do not immediately involve shaping urban form, the interdependencies across city metabolisms from bodies, to infrastructure to the city collective mean that removal from one location (the body or the sewer) potentially leads to a problem of redisposition in another context. Overall, this strategy presents a partial understanding of the metabolism of fat and fails to connect the problem with wider questions about the specific contexts and environments with which bodies and sewers are located. There are serious concerns about its longer term effectiveness and viability as a strategy for dealing with fat.

Preventing the Deposition of Fat

If the removal of fat is complex and expensive, an alternative strategy is that of preventing the deposition of fat. It is not just those with fat bodies who must be anxious about fat, it follows, but as an epidemic, as with any other disease, the public must be aware and enrolled into fitness and diet regimes, into lifelong healthy lifestyles. And this is also the case with the city where there is a potential window of opportunity to maintain health, fitness, and leanness before the crisis strikes. The *Men's Fitness* campaign has stimulated the mobilization of a renewed interesting in the "retooling of the urban environment" to mobilize bodies. The strategy becomes one oriented around the maintenance and acceleration of flows to ensure that the fat remains mobile and that the opportunities for its deposition are limited. In relation to sewers there has been the emergence of a range of promotional strategies, guidelines, and ordinances all designed to reduce the deposition of fat and with that a concern to mobilize other waste infrastructures that could provide alternative routes for fat disposal. And, just as bodies trying to lose fat require the enrollment of a wide set of networked relations from food-supply chains to transport infrastructures that become mobilized through the city collective, so too we find that strategies of removal of fat in sewers involve a vast array of food-supply chains and transport infrastructures, ranging from the local configuration of waste management to the global markets for oils. When we turn to prevention strategies we see again the complex of interdependencies between bodies, infrastructure, and cities, but now we also find the intentional reshaping of urban form. In this sense the need to maintain flow means that to mobilize fat the city needs to become more malleable both physically and in terms of social practices. Not only is the urban environment being actively explored in relation to its implications for the movement and feeding of bodies, but so too, the habits of bodies in houses and restaurants are being reshaped to protect the health of the sewers.

Intervention is much more contextualized, looking at the bodies and sewers and cities in terms of their mutually shaping relations. The body has to be looked at in the context of its movement and mobility through the city; the sewer has to be looked at in relation to the practices of waste disposal by households and restaurants. The role of the city in creating a context for collective action to mobilize bodies and action around the sewers becomes significant. This strategy involves a much more complex set of partners, social interests, technologies, and social practices. But significantly it asks wider questions about the metabolism of fats. Rather than focus on issues of deposition it raises wider questions about bodies in the contexts of what is consumed and eaten and how bodies move through cities. This in turns starts to raise wider questions about the consumption of foods and the distribution of fats. A wider metabolism of fat comes into view. But

what is also significant is that the interconnections between the different metabolisms are brought into view much more powerfully in this approach. This response places bodies, sewers, and the cities in relation to each other. The context needs to be reshaped to provide a context for bodies and sewers to avoid fat deposition.

From "Fat to Fuel"

The final strategy involves the recommodification of waste fats, oils, and grease as a biofuel. Here a waste substance is reused as a commercial resource in a wider strategy of biofuel development of the automobile system driven by a variety of pressures including peak oil, energy security, and environmental issues. This strategy, carefully developed through the activities of activists, community groups, city authorities, environmental regulators, and commercial companies—provides a new metabolic cycle for the urban fats, oils, and grease that seeks to create "circular" or "closed-loop" metabolisms. But of course these are not entirely closed systems because fats and oils still need to enter the city through the food distribution process before they are utilized in restaurants and fast food takeout. But this strategy has some interesting implications in its relations with the metabolism of fats through people and sewers. Critically one of the key justifications for the use of biodiesel is the impact on low income communities, so, for example, the EPA argues that in the San Francisco biofuel initiative:

> Replacing traditional diesel with biodiesel fuel is especially important in communities like the Bayview Hunters Point neighborhood which is heavily burdened by truck traffic on its streets. Use of biodiesel will enable the city to achieve significant reductions in very fine soot particles. These microscopic particles are one of many toxic pollutants that are emitted from diesel exhaust, which has been linked to an array of serious health problems, such as

asthma and premature death. Biodiesel usage also greatly reduces carbon monoxide, hydrocarbons, toxic air pollutants, and greenhouse gas emissions.[393]

Yet these low income communities may also be a critical source for waste fats through collections from fast food and takeout restaurants. Recent research has started to examine the role of obesogenic environments—"the sum of influences that the surroundings, opportunities, or conditions of life have on promoting obesity in individuals or populations."[394] The relationship between obesity and the distribution of fast food and takeout restaurants is one of the many relations being considered within work on obesogenic environments. But these other potential connections between fats and obesity become hidden as the health benefits of waste derived biofuels are compared favorably against the problems associated with traditional diesel fuels.

As demand for waste derived biofuels has increased so has demand for waste oils and fats from restaurants and takeout locations. Initially wastes were collected for free from restaurants and the collection service was paid by the biodiesel manufacturers, but there is now sufficiently high demand for waste fats that it can be sold as a resource by the restaurants. Consequently, the price for waste fats has increased leading to further experimentation in the exploitation of other types of food wastes and animal carcasses that can be converted into biofuels. This even includes the use of human fat extracted through liposuction as a fuel. In 2006 it was reported that a business in Miami was in negotiations with a hospital to give the company about 3,000 gallons of human fat a week from liposuction operations which it was claimed were enough to produce about 2,600 gallons of biodiesel fuel.[395] A Californian plastic surgeon claimed to have turned fat, extricated in liposuction, into biofuel for his cars on his now closed Web site (lipod-

iesel.com) and is under investigation by the California Department of Public Health because it is illegal in the state to use human medical waste to power vehicles.[396] Consequently, we can see that strategies of converting fat to fuel at least partially cross over into human metabolism as the feasibility of using liposuction fat as a resource is apparently further investigated and exploited.

By looking across the three potential responses to fat deposition in sewers, cities, and bodies we highlight powerful affinities across metabolisms between the discourses, social practices, and technologies involved in remobilizing fat. Further still, each of the metabolisms that mobilize fat in one context requires a mobilization in another. People have to be enrolled in terms of intercepting grease and fat from their homes or from restaurants, to keep the sewers mobile; people need to be mobilized to keep fat more mobile within an urban context; and cities themselves need to become retooled or reconfigured to keep the people moving and by converting fat to fuel. Finally each of the strategies embodies a quite different view of the cities' reconfigured relations with fat. A strategy of removal implies that the fat is attacked to ensure that bodies, sewers, and the city can be slimmed down. The strategy of prevention of fat deposition implies a defensive approach in which bodies, sewers, and the city are sociotechnically reengineered to ensure that fat is kept on the move. A strategy of fat reuse involves a strategy of re-co modifying as waste fuel that by bypassing the sewer system can then be used in the automobility system as a resource.

CONCLUSION

The metabolisms of fat in bodies, cities, and sewers are clearly interwoven within each other yet these linkages are poorly understood. The challenge is to understand the mutually defining relations between bodies, cities, and sewers. Evidently the city is a key factor in the production and reproduction of bodies and sewers. The city provides an emerging context and coordination for bodies and sewers, the order and organization that link bodies and sewers, and the framework in which fat is consumed and deposited in bodies and sewers. But the relations between these metabolisms are extremely complex. Bodies and sewers must be considered active in the production and reproduction of the city. These mutually redefining relations can be viewed as series of interfaces between metabolisms. Rather than seeing the metabolism of cities, bodies, and sewers as single entities they should be regarded as assemblages with fat capable of crossing the boundaries between metabolisms to form particular sets of linkages. This is not to stress the unity of an idealized ecological balance but a set of interrelationships that involve a series of flows that are brought together and drawn apart in a series of temporary alignments. The aim of this chapter has not been to present a comprehensive account of the flows of fat across the metabolisms of bodies, cities, and infrastructures. Rather, our approach has sought to highlight the possibilities for following fat in ways that raise the visibility of the interdependencies between bodies, city, and infrastructure within a mobile society and which problematize attempts to isolate interventions to particular sets of relationships without the disruption of others.

Researching the mobile society, our exploratory investigation into fat suggests, requires a close examination of the interconnectivities constituted in mobile society that become more acutely revealed during times of "crisis." Nonetheless, our initial insight into the travels of fat through the metabolisms of bodies, cities, and sewers demonstrates the significance of developing social scientific understanding of mobility beyond people into the material worlds.[397] Understanding the flows of fat in cities requires understanding the complex configurations of bodies, cities, and infrastructure within which fat emerges, transforms, and, sometimes, moves on. Fat presents a case for opening up the social sciences to understanding mobility in terms of the interconnectivity of phenomena and spaces, not in a holistic sense as if all the relationships could be understood, but in terms of a sensitivity to moments of interconnectivity between the interfaces of different metabolisms.

CHAPTER 7

Securitizing Networked Flows: Infectious Diseases and Airports

S. Harris Ali and Roger Keil

This is where an outbreak would probably hit, where the international airports are.[398a]

Monumental terminals of glass and steel designed by celebrity architects, gigantic planes, contested runway developments, flights massively cheaper than surface travel, new systems of "security," endless queues—these are the new global order, points of entry into a world of apparent hyper mobility, time-space compression and distanciation, and the contested placing of people, cities and societies upon the global map.[398b]

One of the key features of globalization, regardless of how it is defined, involves the increased and intensified level of connectivity between diverse sites across the world. It must be kept in mind however, that such connectivity is predicated upon flows that essentially give material form to the interconnections between those sites. Thus as Callon and Law observe, "The notion of connection is not enough. Something has to circulate too. There has to be movement between points of action at a distance for mobilization to be possible. If one place is to be 'globalized' then it has to be linked to others."[399] These connecting flows can take many different forms, as: commodities, information bytes, ideas, capital, labor, and as will be the focus of this chapter: pathogens, people, and airplanes. As many of the chapters in this volume attest, the movement of such flows can only be main-

tained through the establishment of physical infrastructures that are consciously designed to facilitate their movement from one node to the next through various networked systems (e.g., sewage systems, the electrical grid, communication satellite networks, roadways, and so on). The disruption of flows in any of these networks dramatically reminds us of our inherent dependence on these networked systems as we are forced to deal with the unexpected and sudden disruptions of everyday routine practices that arise in the wake of interrupted flows. Network failure often leads to a great deal of public scrutiny as demands are made to discover the "cause" of the interruption of flow. As a result, various government-sponsored investigations are commissioned, but more often than not, these tend to focus on technical matters or operator failure rather than on the organizational and political context within which the (dis)functioning networked infrastructure was embedded. However, the social context is very important to consider because as we will see in the case of the global outbreak of Severe Acute Respiratory Syndrome (SARS), the ability to mobilize resources to address network failures can be either inhibited or facilitated by these very factors.

It should be noted at the onset that one peculiar aspect of analyzing the flow of pathogens is that the disruption to society occurs because of the *continuance* of flow. Thus,

Figure 7.1 Warning sign in Mass Transit Railway (MTR) station in Hong Kong, 2006. Source: Photograph by Roger Keil.

efforts are not directed toward the restoration or resumption of flow, as would be the case in restoring electricity or sewage flow when an infrastructure network ceases to function properly, rather, in the specific case of pathogens, the idea is to break the chain of transmission; that is, to disrupt the viral flow so that outbreak can come to a conclusion. At the same time, however, the flow of "healthy" individuals and goods should be maintained as much as possible so as to ensure the other necessary activities required of a functioning society and economy are retained. As will be discussed throughout this

chapter, the pressing and ongoing practical challenge during a disease outbreak situation is to strike a balance between halting one type of flow while permitting another type of flow. An effective strategy in this light will need to be taken into account the fact that these two flows (i.e., viruses and people) are intimately intertwined. That is, quarantine and isolation are used to halt the transmission of the virus, but such actions should not halt the mobility of the uninfected. Because of the peculiar biological characteristics of the SARS corona virus, particularly its viral reproduction rate, to stop an outbreak of this disease, public health officials needed only to block viral transmission in about half the infected cases. Notably, this is very different from the case of pandemic flu (e.g., Flu A or H1N1) where an almost 100 percent containment rate will be required for the effective disruption of viral flow.[400]

Efforts taken to interrupt the flow of pathogens do themselves rely on the functioning of various networks, particularly those dedicated to disease surveillance. For example, the data concerning case incidence and secondary contacts gained through epidemiological surveillance practices, such as contact tracing, are usually entered into a computerized databank and shared amongst a network of local, regional, national, and supranational public health officials responding to the outbreak at different levels. These epidemiological data are collected at certain nodal points in the institutional network of society, such as hospitals and airports. Notably, these nodal points also serve as sites toward which outbreak containment actions such as quarantine and isolation are directed. The role of hospitals in interrupting the flow of the SARS coronavirus has been considered elsewhere,[401] we will therefore focus here on airports.

THE GLOBAL EPIDEMIC OF SARS

The outbreaks of SARS between November 2002 and July 2003 resulted in the infection of 8,100 individuals and 800 deaths worldwide.[402] The SARS epidemic was unique and notable in several respects. First, it was often referred to as the first infectious disease epidemic of the "global era" because of the rapidity at which the virus traveled around the world. Second, on a related note, the epidemic was the first of its kind in terms of the extent to which airports and airlines were instrumental in the spread of the disease.[403] Third, it was noteworthy that the outbreaks

Figure 7.2 Temperature screening post, Port of Hong Kong, returning from Macau, 2006. Source: Photograph by Roger Keil.

did not occur in cities of the Global South (where it was thought the most serious of infectious disease outbreaks would first establish themselves), rather the virus surfaced in some of the most developed and advanced cities of the world, that is, in global cities such as in Beijing, Hong Kong, Toronto, and Singapore.[404] Finally, the technical response to the epidemic was exceptional. Within a few weeks of the initial outbreaks in global cities, a virtual network of scientists and public health specialists was established through the coordinating efforts of the World Health Organization (WHO). Temporarily casting aside competitive aspirations, epidemiological, virological, and clinical data were shared by these scientists. The collaborative efforts led to the successful identification of the causal agent of SARS and the subsequent characterization of its genetic code, in the record time of two months.[405]

AIR TRAVEL AND THE POLITICS OF PUBLIC HEALTH

The political issues related to infectious disease spread have a long history that is closely associated with the development and institutionalization of modern public health and international public health diplomacy. The SARS outbreaks have revealed that many of the same issues that have plagued historical attempts to contain disease spread via shipping in the time of mercantile capitalism remain in today's age of jet travel. Perhaps the most significant and persistent of these issues involve the tensions between public health and international trade. As early as 1851 when the first International Sanitary Conference was held by European states to discuss cooperation in addressing the trans-boundary transmission of cholera, plague, and yellow fever, the preferred method of response was quarantine.[406] However, state-imposed quarantine practices interfered with the ability of nation- states to engage in trade, thus violating the internationally agreed upon Westphalian principle of minimum state interference with international travel and trade. The response to the SARS outbreaks by the WHO blatantly violated this long-established political principle, as it "regard[ed] every country with an international airport, or bordering an area having recent local transmission, as a potential risk for an outbreak".[407] And on this basis, the transnational public health agency publicly recommended (for the first time on April 2, 2003)—through various press announcements and postings on their Internet site—that potential international travelers postpone all but emergency travel to areas of local SARS transmission. By taking such action, the WHO was essentially influencing the course of international trade and travel. Such "interference" by this supranational organization was unprecedented and the economic impacts of the WHO travel advisory were significant, especially in SARS-affected cities that had to deal with the cancellation of conventions, the drastic decline in tourist travel, and empty inbound airplanes.

The WHO also "recommended" that airports in SARS-affected areas adopt certain surveillance practices including: temperature screening of departing and transiting passengers, the provision of information leaflets to travelers, exit questioning, and the completion of a mandatory health declaration form by passengers.[408] The issuing of these recommendations violated another principle of the long-held Westphalian political order, namely the principle of state sovereignty; that is, that the nation-state holds the exclusive right to govern domestically in an autonomous manner free from external influence. Although these recommendations appeared as voluntary, in effect they were not because if the nation-state did not adopt the WHO recommendations, then the travel advisories would remain in place, thus having a continued, and undesirable economic impact on SARS-affected cities. The issuance and continuance of the travel advisory therefore represented ways of ensuring compliance of the nation-state to WHO demands, and in this connection it was noted by one infectious disease specialist involved in the Toronto SARS response that:

> The WHO had suggested airport screening, this had not been implemented, there was a whole bunch of rumours going around. And it was like the WHO was really pissed off with Canada and then, this was sort of how we got our hand slapped, with the travel ban.[409]

In late April 2003, shortly after the issuance of the travel advisory against Toronto, a Canadian delegation consisting of city, provincial, and federal politicians and health officials visited the WHO headquarters in Geneva to persuade WHO officials to lift the travel advisory. The advisory was lifted under the condition that infrared thermal scanning of passengers would be implemented at international airports in Canada. Several days after the lifting of the travel ban, a second significant outbreak of SARS occurred at a Toronto hospital leading critics to contend that due to political and economic pressures there was a premature lessening of public health vigilance, as exemplified by the lifting of the travel advisory.[410] Others, argued however, that the threat of imposing the travel advisory represented an obstacle to taking effective public health measures because it meant that cities and nation-states may be less willing or forthcoming in revealing information about potential outbreak situations for fear of potential economic loss.[411] Such discussion led to the questioning of the general effectiveness of the travel ban and airport screening as strategies of outbreak control more generally. Second, it raised the related issue of the significance of airplanes and airports as sites for disease transmission. By the end of August 2003, an estimated 6.5 million screening transactions had occurred at Canadian airports with 9,100 passengers referred for further assessment by screening nurses or quarantine officers.[412] Of these, none were found to have SARS. Other countries yielded similarly low results.[413]

SECURING THE AIRPORT INFRASTRUCTURE

"Immobile" platforms such as roads, garages, stations, docks and so on *structure* our mobility experiences.[414] We rely upon these immobile platforms and the networked infrastructure in which they are embedded to support our daily activities, and we often do so in an unquestioning and taken-for-granted manner. Such an outlook may be the outcome of the fact that many infrastructures exist outside of our regular viewscape, such as under the ground, as in the case of sewage and water pipes, or high above the ground, as in the case of power lines.[415] Nevertheless, the existence of these latent infrastructures

can significantly influence the movement and activities of individuals, often, in unsuspecting ways. In this light, Doganis notes that unbeknownst to travelers, the physical and aesthetic layout of airports is consciously designed to exert control over the movement of people in predetermined directions through the use of a multitude of strategies.[416] For example, the physical construction of terminal buildings, including its corridors and walls, is intended to limit the possibilities of movement and action so as to ensure "proper" and "correct" movement through various spaces such as at the check-in counter, the security control checkpoint, the departure waiting area, the boarding corridor, and interestingly, to channel people through the commercial areas of the airport to ensure that they receive adequate exposure to the wares being sold. This is done in a seamless manner in which the passenger is faced with the options of moving only forward or backward thereby creating "an environment that invites an automatic response from the passenger; those who have not been to the airport before intuit their projected path according to their situation."[417] The airport is also designed to alter the emotional state of passengers so as to discourage personal interaction by creating and maintaining an atmosphere of formality that respects the seriousness of the work of securing the airport:

> [I]t is thought by creating an uninteresting, and quite oppressive security environment, the idea in many airports has been to induce feelings of melancholy and, to an extent pressure. They do this in the hope of limiting what people do in these spaces....[T]he emotional state of the passenger—affected by the airport environment—is meant to literally close-off the passenger's capacity to disrupt the security processing system through, for example, walking the wrong way, or by telling a joke or misbehaving.[418]

In this connection, Mark Gottdiener has made a related point very convincingly in his treatment of airports: "The airport has taken on the characteristics of Simmel's city to an extreme. It has all the trappings of a thoroughly instrumental space with even less of a need for people to interact. In fact, the airport norm is one of *non-interaction*."[419] Thus, airports are characterized by some as "nonplaces," where people coexist or cohabit without living together, in essence creating "solitary contractuality."[420] Indeed, the production of this type of place of social indifference and magnified civil inattention may be characteristic to other sites that act as points of convergence for global flows, such as hotels. This is seen for example in Sofia Coppola's film *Lost In Translation*, where "global" actors (or better, Americans) are desperately trying to find humanity in a Japanese city that is commercialized and alienated to the extreme, thus dramatically depicting the tensions that exist between global network space and human (re)pro-

duction characteristic of global flow convergence nodes.

Callon and Law ask the general question: How is security and order maintained when people and things are constantly shifting positions?[421] That is, how can order be produced by managing the multitude of interacting flows? Answers to this can be gleaned by considering the many strategies of social control employed at the airport. By passing through security points, possible threats are meant to be filtered out, thus "resulting in the 'sterilized' passenger who may enter the 'sterile' zone of the duty-free airside concourse.[422] Furthermore, since the airport represents a point of contact between the individual and the state,[423] it is also a site where power relations are enacted, as perhaps best illustrated through the common experience of being forced through bottlenecks where "those in the corridors of power may exert influence over those in the

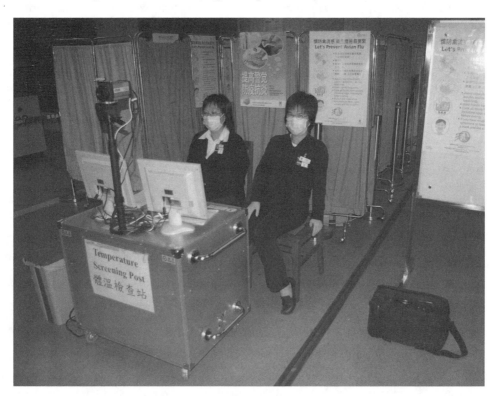

Figure 7.3 Quarantine Station in Narita International Airport, Tokyo, 2008 Source: Judith Versloot, by permission..

corridors of movement."[424] In this context, airport checkpoints are "transfer points" or "places of in-between-ness" where:

> "populations" who are mobile can be monitored by various agencies charged with policing that territory; and simultaneously can be researched since they are temporarily immobilized—within lounges, waiting rooms, cafes, amusement arcades, parks, hotels, airports, stations, motels, harbours and so on. These transfer points necessitate a significant immobile network that is partly concerned to effect surveillance of intermittently moving populations.[425]

As Foucault has pointed out, the sorting and resorting of populations through tactics of surveillance is fundamentally a political question, implicating the use of power and knowledge in the exercise of social control.[426] Since the airport represents a site of considerable sorting and resorting of disparate populations across national borders,[427] it is no surprise that surveillance plays a critical role in contemporary airport operation. Clearly in the post-911 era, airport surveillance in the service of "security" has intensified as indicated by the introduction of ever more pervasive surveillance technologies,[428] as seen, for example, by the increased use of detention centers, closed-circuit television cameras (CCTV), global positioning systems (GPS), iris-recognition security, intermodal traffic interchanges,[429] as well as through the embedding of data codes on airline tickets that digitally enscribe what the passenger is doing and predict other actions that the passenger may take.[430] Still further indications of the extent to which the post 911 "state of emergency," or the exception has become the "rule" is to consider just a few of the many recent state initiatives aimed at "securing" the airport. One example involves the "nationalization" of airport security with the Department of Homeland Security in charge of implementing standardized systems of person and baggage securitization.[431] A second example, is the U.S. government's Total Information Awareness program that integrates and coordinates different types of data from private and public sources, including those from biometric technologies that are able to recognize humans at a distance and the mapping of people's multiple connections across their social networks.[432] While a third example is given by the passenger profiling systems known as Computer Assisted Passenger Prescreening or CAPPS (now Secure Flight), that works to gather as much information about a passenger with the ostensible aim of enabling officials to make informed judgments and risk measurements about the passenger's propensity to become a threat.[433]

These calculative surveillance systems designed to gauge the individual's potential as a terrorist threat have their parallel in relation to the discernment of a public health threat that takes the form of the infecting other. It is in the context of the latter, that attention to airport security has increasingly been focused upon the development and implementation of strategies of thermal screening, questioning of health status, detention, quarantine, and isolation as part of an overall paradigm of "public health security." The Quarantine Act, for instance, allows Health Canada's quarantine officers to detain individuals in order to conduct medical exams. The Act also gives the government the right to impose the requirement that airlines distribute health information and conduct active screening of travelers. During the SARS epidemic, the quarantine responsibilities were transferred from Health Canada to the Canada Customs and Revenue Agency; however, it was noted that the customs staff were never trained for these duties and airport authorities expressed concerns about this, citing inadequacies in the provision of logistical support.[434]

THE SECURITIZATION OF SPACE

Spatial considerations are central to both national security and public health security

TOTAL DEFENCE
For better or worse

Singapore's ring of defence is one of its key pillars, and has kept its people safe from threats like terrorism and SARS. As citizens, we have a part to play in our country's total defence. We need to be prepared to protect ourselves from harm and stay vigilant to keep 'the enemy' at bay, come what may.

Figiure 7.4 Slogan in public exhibition to celebrate national unity in Singapore, 2006. Source: Photograph by Roger Keil.

initiatives at the airport, as both domains of activity require a physical space in which to operate, especially in relation to the detention of those under "suspicion." In fact, according to Sarasin, concerns about terrorist threats and infectious others may have become ideologically conflated in public consciousness and public discourse, to such an extent as to even serve as the legitimizing basis of the U.S. invasion of Iraq in 2003.[435] Fighting emergencies creates very clear lines of demarcation between more or less secured spaces. The Baghdad "Green Zone" is perhaps globally the most well-known example of a militarily secured space. In global cities, the most secured areas are, as a matter of course, the financial industry's vaults and computer networks, the corporate headquarters, and other network nodes where the flows of global capital and information are moored, where surplus value is extracted from worldwide production and supply chains. The upper level producer services that are concentrated in the high rise office buildings of downtown are locked down in electronic and human controlled security systems, in a built environment of glass and steel seemingly impervious to the intrusion of unwanted and unexpected entries.

The moored security of the "citadel" stands in clear contradiction to the insatiable demand for mobility and flexibility of spaces created by global capital, hence the quandary faced in imposing airport security measures.

Spatial fixes are having a short lifespan these days, and the mantra of mobility rules the everyday of the senior operatives of global capital who are employed in secured inner city spaces. While those operatives and their work need to be safe here, they also need to be secure over there, in the network which connects them to other nodes in the web. This dialectic of safety on one hand and the need to leave the cocooned spaces of the secured "citadel" on the other is a constant challenge to the architecture of the global corporation as it is to the contemporary airport. It necessitates the construction of supply chains of safety across the globe, in which high speed trains, airports, lounges, hotels, clubs, high end entertainment districts, and the like are knitted together into a landscape which extends the global regime of accumulation across the globe. The airport gains iconic status in this supply chain of corporate safety and it is here where the security interests of the state and the desire for safety of economic transaction on the side of the corporations coincide most visibly.

The emergence of "unbound" urban regions involves the sectionalizing of territory for infrastructural and other specializations. Thus, airports, for example, have traditionally been built on the edge of cities as places or camps of banishment,[436] while today they serve as indispensable gateways to the global city. Indeed, Urry notes that airports are one of the key ways in which cities seek to enter or position themselves within the contemporary global order.[437] An indication of the importance given to this is seen by the fact that increasingly, cities are competing to build the largest, newest, most expensive or stylish airport.[438] The airport itself is, of course, just a dot in the networked and webbed landscape, the actor network of international travel. Latour famously said: "Boeing 747s do not fly, airlines fly."[439] Airlines are, in fact, already partners in the work that the WHO does and SARS did not change much about that:

> Well, I mean in…airports…we certainly communicate with ICAO in Montreal, the International Civil Aviation Organization who perhaps have something to do with airports, we also deal with IATA, who deal with airlines, and they are certainly very interested in…and flu pandemics and… So we would certainly have relations with different sectors. But I mean as far as is there anything different regarding administrative levels that we would…then I'm not aware of anything that is particularly new in that as the result of SARS.[440]

As indicated by the instances reviewed above, during SARS, airports became major points of disease management and control. Helped by the "new normal" after 9/11, airports became the technological interfaces between the global (imagined as threatening) and the local (imagined as safe). During the second outbreak in Toronto, after the city had been hit with the WHO's travel advisory, the relationship was inverted as the airport became the control points for measures to protect the local from the global. Temperature screening stations, information campaigns and other medically based measures implemented at that time (as reviewed above) changed the symbolic meaning of the airport as a place of global mobility to a place of local restriction. In Toronto, the lessons learned from SARS were swiftly boiled down into a new policy that reinforced the autonomous status of the Greater Toronto Airport Authority by placing it at the core of the

governance of the pandemic as they would be charged with coordinating the airport-related stakeholders in the private sector, the public sector, and the security community.[441] To some degree this meant that in the case of a future pandemic, the unboundedness of the city would be compromised through the gatekeeper measures ascribed to the airport authority, which would mean a decentering of power both in an institutional and in a geographic sense. A policy drafted by the GTAA in response to SARS spoke first of "sterilizing passenger flow" and the separation of the running of the airport from the potential threat that passengers pose. This involved a specific management of facilities in order to minimize contact. Second, with regard to the overall process of infectious disease management at airports, the GTAA saw its role as one of coordination and facilitation in order to let the health specialists do their job effectively (although this may not have been successful according to the comments of one local public health official—see below). And third, the airport recognized its role "as part of a global network."[442] This peculiar mix of place-specific and network-directed initiatives is, then, *typical* of the role of urban regions in the management of global infectious disease as they combine territorial with topological strategies of network survival.

Airports, however, are not going to be replacing the health care institutions elsewhere in the urban region, although SARS raised the importance of establishing a more formalized relationship between the two, especially in management of space that is thought necessary for the segregation of those infected. For example, in Hong Kong, a senior medical officer responsible for the surveillance of infectious diseases noted that:

> There's no containment camp in the airport. Depends on the scale of the containment site you're talking about. For the airport, there is a special room, if they detect some infectious disease case, an isolation room. They will keep

them there and transfer the patients to our hospital immediately for further management and investigation.[443]

Similarly in Toronto, a particular hospital was selected to specifically deal with SARS cases because of its proximity to the airport.[444] Indeed, this hospital was praised for its strategy of employing the innovative "Hospital Within a Hospital" model that enabled it to maintain its operations as a fully functional health center while at the same time operating a self-contained parallel facility to deal with SARS patients.[445] Such was one strategy of dealing with issues of "boundedness" yet "permeability," a strategy based on maintaining mobility concurrent with detention—an objective that airports also sought to emulate to ensure their nodal role in facilitating the movements required of local and global capitalism.

No SARS patients were treated at any airport as far as we know, with the exception, perhaps, of a Frankfurt case which saw on-site medical attention given to a Singapore doctor, infected with SARS, and his travel companions, before they were isolated in the city's university hospital. Frankfurt, like other international airports, is classified as a "sanitary airport" fulfilling specific stipulations of the World Health Organization's International Health Regulations including the existence of medical organization, personnel and space, the capacity to transport and isolate travelers who are potentially infected, disinfection infrastructure, a bacteriological laboratory, and inoculation capacities. In this case, under the direction of the City of Frankfurt's public health department, the airport figures prominently in a network of "competences" designed to battle highly infectious, life-threatening diseases. This network also includes the University hospital and the fire department.[446] In the same vein, the specific geography of Pearson International Airport in Toronto—which has a large infield terminal used during construction of its new Terminal 1—allows for the possibility of "ideal isolation" to which passengers from abroad can be brought and kept from other areas of the airport if need be.[447] The airport in general, and the infield terminal in particular, have the advantage, of course, of being a far distance from most residential and other urban uses. No natural contact can occur. Airport security, which provides an important, perhaps the central cordon sanitaire between the outside world and the city (or the other way around) revolves historically around "contraband and terrorism." While "Health Canada has always been a part of the airport," the increased worry about infectious disease has now been added to the ways in which the airport secures the urban region: "Before the idea was to get people to a safe, secure area. Now it's to get into a sterile area. And sterile is in terms of hygiene, not in terms of security."[448] In sum, medical emergencies have now been added to the "normal" process of security arrangements as an "operational procedure. We have an aircraft coming in, health wise, and here's what we've got to do. And again…the requirement is quite different from security or emergency."[449]

In a globalized environment of generalized risk, metropolitan airports are important biopolitical locales: Because the New Orleans airport became literally a hospital in the wake of Hurricane Katrina in 2005, the manager of that airport debriefed colleagues at other airports. The increased vulnerability of local places in a case of globally induced emergency leads to counterintuitive place-specific restraints at exactly those places that usually symbolize global connectivity. This "splintering" of network functions is typical of today's networked urban world.[450] After SARS and Hurricane Katrina, for example, the Toronto airport built a cogeneration plant to increase its independence from the power grid. Already equipped with "emergency services in terms of bottles and blankets and all that stuff," the airport now is reimagined "as a bit of a resource for a community…. We have large buildings, and we have a lot of facilities that could become available to a community. Especially if there was a pandemic that required…sort of isolation." The airport would become a "staging area for emergency services for the army."[451]

These kinds of tasks that involve the airport as a proto-military or security site also link into the functions the airport performs to keep air traffic running and to link this goal to the public health objective to "keep the pandemic outside." Various screening and reporting mechanisms present at airports are tied to specific emergency plans which turn the airport's infrastructure of mobility (ground transportation, buildings, checkpoints, etc.) into a landscape of medical control, quarantine, and surveillance.[452]

Just as the response to SARS was complicated by differing and competing jurisdictional responsibilities between the WHO and nation-states, similar difficulties can be discerned when considering the relationship between the autonomously functioning Greater Toronto Airport Authority (GTAA), the federal government, and the local public health unit. Traditionally a federal responsibility, international air transport fell under the jurisdiction of Transport Canada, but as another example of the implementation of neoliberal policies in Canada, the private agency of the GTAA assumed the management, operation, and maintenance of Toronto Pearson International Airport from the government on December 2, 1996.[453] Complications arose and were brought to the fore during SARS because Toronto Pearson International Airport is actually physically situated outside the City of Toronto in a neighboring municipality. As such, technically the local board of health of this adjoining suburban municipality is responsible for the public health of all individuals within its boundary, including those at the airport. Further complicating things is that the municipality in Canada is legally a corporation that falls under provincial jurisdiction. These circumstances have

led to a strained relationship between the local board of health and the Greater Toronto Airport Authority. As glaringly evident from the statements of one senior public health official in this municipality who expressed the view that the GTAA was not receptive to input from the local public health unit:

> Because the airport has been run by the GTAA, they're just doing whatever they want and they don't care because they are falling through the gaps by the feds, the province doesn't pay attention because they don't have enough time and resources to dedicate to their own programs, and here we have a population [in the local municipality] that is potentially at risk because all of these things have not been followed up on.[454]

> They just want to do their own thing. And it is not just food and tobacco and DSRs [Designated Smoking Rooms], it's also issues of emergency response. We weren't included in their emergency exercises. I said "how can you leave us out? we are…you guys reside within [the local municipality] even though you're federal lands. What you guys do will affect people." Like this Air France incident, all the chemical spewage that came out of the plane— on the cargo and the actual plastics on the plane—it went into one of the rivers and that river goes into Lake Ontario, but in the process of getting to Lake Ontario it goes through residential areas, schools, other areas where kids may be playing or close to the water. How can you ignore us? And there was oil spills, a whole bunch of stuff. Every health emergency that happens federally affects us locally as well. Because we have the airport. And they've essentially said, "No, we don't care and go lead your own life.[455]

The jurisdictional squabbles and lack of cooperation between different scales and in particular across the public–private divide clearly reveals how the "splintering" effects[456] of privatization, vis-à-vis the reduced ability of the state to govern infrastructures such as those associated with airports and public health, make certain places particularly vulnerable to transboundary environmental and health threats such as disease outbreaks, particularly in light of *how* the world is interconnected today—the subject with which we conclude.

CONCLUDING REMARKS: TECHNICAL, SOCIAL AND ECOLOGICAL CONNECTIVITY, AND THE SECURITIZATION OF FLOW

With the convergence of accelerated polyrhythmic flows of different types, the city takes on a networked, yet unbounded and emergent quality, where the urban is always in a state of flux as driven by the constant movement of diverse flows through it.[457] These dynamic qualities of the modern city have made it difficult for local authorities to deal with mobile threats,[458] whether they be in the form of terrorists or viruses. These problems are further exacerbated because these very same changes in the dynamism of the city have triggered changes in the politics of the global city, particularly in the ascendance of those involving a crisis politics of the "new normal" based on a regime of heightened vigilance and suspicion.[459] Most experts, of course, rate the current danger of bioterrorism as a real but unlikely threat.[460] But its paradigmatic importance has led to putting airport agencies on alert and has made airports into frontlines of national and international biosecurity practices.[461] Such developments, both in terms of mobility and politics, have changed the map of vulnerability of the city, and one of the effects of this is that the airport has become a critical site to address the fallout and problems that arise under such conditions, such as the global spread of disease. The rather unpredictable effects and the resulting vulnerability that arises may be understood in terms of connectivity, or more specifically, the convergence of different types of connectivities—technological, social, and ecological.

The physical infrastructure of the airport is clearly a technological project, but its function is to perform a diverse number of tasks that include those related to the physical act of flying (e.g., the maintenance of runways, air traffic control, etc.), or the information and communications systems required for surveillance. In all cases, technological functions support some social objective, whether it be the needs of travelers or the needs of security. In this sense, technological connectivity and social connectivity must be seen in relation to one another. For our purposes here the spread of SARS in a globally connected world is a consequence of this increased technical cum social connectivity, because, as we have discussed above, the global flow of the virus could not have occurred without technological means such as the airplane, nor could its containment occur without an established communications and information infrastructure of the local and global public health systems.

But, something else is at work here in studying the relationship between infectious disease spread and air travel. As Harvey has observed so insightfully, there has been a rescaling of the connection of the body to processes of globalization.[462] This insight casts light on those processes and topologies that may be commonly underrepresented in depictions of globalization. When global travelers leave the "deterritorialized" spaces of airports, hotels, and first-class lounges and where people interact in the interstices of the globally connected money markets of the global city, places and spaces come into sharp relief, where the production processes of globalization really occur through and in the everydayness of the global city network's multiscaled neighbourhoods.[463] This systemic and networked integration of technosocial practices also relies on the existence of a specific division of labor that undergirds the global mobility of capital and bodies. Whether they are active at the airport or outside its perimeter in the city itself, "the bodies of hospital and hotel workers are a central site of the renegotiation of urban and global security in the face of emerging infectious disease."[464] The circuits of these specific practices of workers intersect, sometimes uneasily, with the hierarchical regulations to combat and preempt disease from the WHO's International Health Regulations to each employer's own pandemic preparedness plans.

In this context, the disease outbreak as the unexpected consequence of flow convergence highlights the importance of place, such as the airport site, as a location where the risks of social and technological interactions of global capitalism manifest themselves in the so-called risk society.[465] At the same time, as Beck explicates based on Aihwa Ong's work on SARS, there are "global assemblages" that both enable and disable infectious disease outbreaks.[466] In the global city system—perhaps one such assemblage—airports play a specifically important role in the "mobility stream" of disease. But to gain a complete picture of how dovetailing connectivities give rise to such risks entails a consideration of another type of connectivity, namely, that informed by biophysically defined ecological processes.

The diffusion of SARS could only occur if human and animal hosts were available and as we have seen, the nature of the diffusion pattern was greatly influenced by the characteristics of the virus as a biophysical entity (e.g., the reproduction time of the virus, its infectivity rate, and incubation period).[467] Medical commentators are making this point forcefully: "Nature itself is the best bioreactor for apocalyptic biological agents and, through evolution, has at the same time developed the best defense strategies against them."[468] Yet, we do know that such processes of disease spread and containment are not just natural but "assemblages" in the sense used by Beck and Ong above and also found in the work of Latour, Callon, and Law.[469] If social communities and individuals get bet-

ter connected through enhanced technological connectivity, while processes of global capitalism lead to new forms of social interactions, such developments in the case of new and emerging disease outbreaks are also shaped by processes of microbial traffic. The term *microbial traffic* refers to the various dimensions involved in the spread of infectious disease, including: (1) the mechanism involved the spatial diffusion of pathogens; (2) pathogenic evolution, including changes in the structure and immunogenicity of earlier pathogens; (3) changes in the human–environment relationship and (4) cross-species transfer.[470] Such characteristics clearly played a role in the spread of SARS.

Ecologically previously unconnected or less connected areas, biotopes, and species are now potentially connected through new forms of sociotechnologically rescaled activities such as air travel. Cross-species transfer (zoonosis), for example, played an important role in the biological origins of the SARS outbreaks. Rural China, where a virus based in animals (i.e., the suspected species being the palm civet cat although there was also evidence of the presence of the virus in the horseshoe bat population)[471] crossed the species boundary into humans, most likely through the handling of animal carcasses, is now only an airplane ride away from distant places on the globe, such as the Toronto region. In turn, a health care worker, who may have become infected in Toronto, is only a plane trip away from a wedding party in the Philippines, where she may infect an entirely unrelated group of people. Further, the infection of a member of a tightly knit Toronto religious community, whose very existence appeared as the epitome of parochialness, had become part of a health crisis of global proportions. These examples illustrate some of the ways in both people and viruses are breaking down traditional boundaries of time, space, and the human everyday. Microbes no longer remain confined to remote ecosystems or rare reservoir species, for them, the earth has truly become a Global Village. These circumstances highlight the importance of taking into account how changes in technological, social, and ecological connectivity contribute to the formation of risks that manifest themselves at particular sites, such as airports, where different global flows converge. Global cities are linked in new networked sociotechnical connectivities. Airports play a strong and inevitable nodal role in setting up these connectivities. Through airports and their ancillary practices, network flows are being securitized. We have discussed in this chapter how the 2003 SARS crisis has accentuated this process through the case of an emerging infectious disease.

CHAPTER 8

Disruption By Design:
Urban Infrastructure and Political Violence

Stephen Graham

If you want to destroy someone nowadays, you go after their infrastructure. You don't have to be a nation state to do it, and if they retain any capacity for retaliation then it's probably better if you're not.[472]

URBAN ACHILLES

In a 24/7, always-on, and intensively "networked" urban society, urbanites, especially those in the advanced industrial world, become so reliant on infrastructural and computerized networked systems that the disruptions become much more than a matter of mere inconvenience. Rather, they creep ever closer to the point where, as Bill Joy puts it, "turning off becomes suicide."[473] Processes of economic globalization, which string out sites of production, research, data entry, consumption, capitalization, and waste disposal across the world, merely add to the "tight coupling" of infrastructure. This is because such systems rely ever-more on complex combinations of logistical, information, and infrastructure systems, working in intimate synchrony, in order to function.

But it is important to remember that the absolute dependence of human life on networked infrastructures exists throughout the modern urban world cities, not just in "high-tech" cities. This has been shown in horrifying detail when states deliberately "de-electrify" whole urban societies as a putative

means—influenced by the latest "air power theories"—of coercing leaders and forcing entire populations into suddenly abandoning resistance. Of course strategic bombing rarely, if ever, has had such an effect. Instead, as we shall see, the effects of urban deelectrification are both more ghastly and more prosaic: the mass death of the young, the weak, the ill, and the old over protracted periods of time, as water systems and sanitation collapse and water-borne diseases run rampant. No wonder that such a strategy has been called a "war on public health" or one of "bomb now, die later."

Urban everyday life everywhere is stalked by the threat of interruption: the blackout, the gridlock, the severed connection, the technical malfunction, the inhibited flow, the network unavailable sign. During such moments, which tend to be fairly normal in cities of the Global South, but much less so in cities of the Global North, the vast edifices of infrastructure become so much useless junk. The everyday life of cities shifts into a massive struggle against darkness, cold, immobility, hunger, the fear of crime and violence, and, if water-borne diseases are a threat, a catastrophic degeneration in public health levels. The perpetual technical flux of modern cities becomes, in a sense, suspended. Improvisation, repair, and finding alternative means of being warm and safe, drinking clean water, eating, moving about, and disposing of wastes quickly become the overriding imperatives of everyday life. Very quickly the normally hidden background infrastructure of urban everyday life becomes, fleetingly, palpably clear to all.

This means that "tremendous lethal capabilities can be created simply by contra-functioning the everyday applications" of many everyday urban infrastructures.[474] The use of systems and technologies that previously tended to be taken for granted, ignored, or viewed as banal underpinnings to everyday urban life, thus become increasingly charged with anxiety. Unknowable risks connected with internationalized geopolitical conflicts are palpably infused into everyday technological artifacts. The post-Cold War landscapes of "asymmetric" conflict blur into a transformation of banal technical artifacts of urban material culture into potential weapons causing death, destruction, and disruption. But states also wield massive power through the threat, or use, of infrastructural disruption. Russia's status as a resurgent power under Putin is due less to its territorial ambitions or military might than the way it continually threatens to, or does, interrupt energy supplies to Southern Asia and Europe, both of which are increasingly reliant on its massive reserves.

Of course, anxieties surrounding risks of infrastructural disruption and destruction are not entirely new. Warfare and political violence have long targeted the technological and ecological support systems of cities. World War II bombing planners evolved complex methods of perpetuating "strategic paralysis" through destroying transport systems, water infrastructures, and electricity and communications grids. Car bombs, of course, have been the staple of every insurgency and terrorist campaign for at least the past four decades.[475] Nevertheless, it is clear that the sophistication with which everyday infrastructures are attacked and exploited to project lethal power is dramatically escalating.

In such a context, this chapter critically explores the relations between urban infrastructure and state and nonstate political violence. Discussion concentrates first on the better known efforts of terrorists and nonstate insurgents to appropriate or disrupt networked infrastructures as means to massively amplify the power of their political violence. The chapter then goes on to discuss the ways in which the military doctrine of nation-states legitimizes and practices the systematic demodernization of entire urbanized societies that are deemed to be adversaries. The use of U.S. air power and Israeli military doctrine against the Occupied Territories are the two most pertinent recent examples. A brief penultimate section discusses emerging ideas of inflicting paralysis on adversary nations and cities through the electronic distribution of malign computer code. The chapter concludes by reflecting on the implications of the centrality of networked demodernization within contemporary practices and theories of war.

INFRASTRUCTURAL TERRORISM

Technology calls into being its own kind of terrorists.[476]

Most large technical systems become extremely uncontrollable dangers if one repurposes their instrumental applications to cause harm rather than [to] create power or profit.[477]

Most attention, so far, has centered on how nonstate actors can dramatically boost their destructive potential by appropriating, or targeting, the embedded systems that sustain modern urban life. As political theorist Tim Luke suggests, "the operational architectures of modern urbanism by their own necessities design, deploy, and dedicate what ironically are tremendous assets for destruction as part and parcel of mobilizing materiel for economic production."[478] In such cases, it is "'technological civilization' that is the target...and the contradiction is that it is this civilization's technology that will be used against it."[479]

The products of Enlightenment designs to overcome the constraints of nature, geography, and time, the achievement of global grids of exchange, ironically, are now open to the designs of terrorists who can manipulate both their global reach and their symbolic power as icons of the global dominance of occidental capitalism. Here we face one aspect of the much-vaunted collapse of grand narratives of Progress that has come with processes of postmodernization. The infrastructures of today's "fast capitalism," in particular, are the ultimate symbols of Westernized global capitalism, whose "politics repudiate fixed territories, sacred spaces, and hard boundaries in favour of unstable flows, the non-places used to stage consumer practices, and permeable borders."[480] Such "Big Systems," though, are open to "asymmetric" violence from nonstate actors who could never hope to counter conventional Western military might.

John Hinkson points out that it is the intersection of these "big systems" and global cities that invariably dominate the targeting strategies of contemporary terrorists. Here he speculates that it is the processes of forming ever-more dominant metropolitan centers that underpins the alienation and radicalization that fuel infrastructural terrorism. "The global city," he writes:

> as it has unfolded during the 20th century, is a great maw sucking in and emptying out regional populations on a scale never seen before. It needs to be understood in relation to the socially thinned-out and increasingly poverty-stricken modes of life in regions, and among certain social sectors, that are stage by stage being turned into dependent, dysfunctional satellites of the metropolitan centres.[481]

The most obvious examples here, of course, were the devastating airline suicide attacks of September 11, 2001 (Figure 8.1). In this example, massive, kamikaze-style cruise missiles were, in effect, fashioned out of just four of

Figure 8.1 The September 11th 2001 attacks on New York (Photograph by Wallyg. Source: http://www.everystockphoto.com/photo.php?imageId=648538

the few thousand of so airliners flying above and between U.S. cities at any one time (or 40,000 or so per day serving around 2 million people). But the attacks relied on a wide range of "means that we associate with Western modernity after globalization, such as money, stock market speculation, computers, aeronautic technologies, and media networks, with the intention to destroy them."[482]

Strategic and symbolic targets at the very metropolitan heartlands of U.S. military and economic power were devastated in the attacks and thousands of people were murdered in a few hours—effects way beyond the power of the entire Nazi or Japanese regimes during the whole of World War II. As the World Trade Center towers collapsed, destructive power reversing the gravitational and architectural hubris of modernist skyscrapers, and approximating in power to a small nuclear bomb, was unleashed. Massive infrastructure failures across large parts of Manhattan and the eastern seaboard were, in turn, manufactured through the use of everyday infrastructures as weapons.

The later Madrid and London train, bus, and subway attacks of 2003 and 2005 are also key examples here, as are the long-standing exploitation of the inevitably crowded "capsular" spaces of Israeli buses by numerous Palestinian suicide attacks to horrifying effect in the Al-Aqsa intifada. In such attacks Palestinians have used one of the few weapons that they have—the continuing ability, albeit in smaller numbers because of tightening restrictions and lengthening fences—to move their (armed) bodies into close proximity to people they assume to be the enemy (Israelis). In martyring their lives in such a way, they wreaked carnage and havoc on contemporary Israel's sites of everyday urban modernity, especially between 2000 and 2002. The chosen targets—cafés, restaurants, bars, buses, bus stops, pool halls, and shopping centers—were selected carefully. In a highly urbanized country such as Israel, they are, by definition, unavoidably crowded places. Israeli response to the attacks has, in turn, stressed that the symbolic urban places of Israeli nationhood—specially coffee houses—are under direct attack. "This is a war about the morning's coffee and croissant," wrote Adi Shveet in the newspaper *Ha'aretz* in March 2002. "It is about the beer in the evening. About our very lives."

Such attacks have raised widespread anxieties about the vulnerabilities of all manner of basic, everyday infrastructures that, by definition, permeate the everyday life of every modern urbanite. Consider here the still-unresolved mailings of anthrax spores, perpetrated through the U.S. postal system in the wake of the 9/11 attacks, killing five people. Or the case of the Washington snipers who turned everyday highways and gas stations in and around the Beltway suburbs of the city into killing fields in October 2002,

murdering ten people. Or consider the fears about the misuse of nuclear material, or the mass poisoning of water or food production systems.[483] Consider, too, the widespread fears that the computerized nature of advanced societies could become their Achilles heel as remote and unknowable "cyber-terrorists" launch malign code into crucial systems, bringing about some type of "electronic Pearl Harbor" in the process.[484]

Nation-states and city governments have responded to these threats through a host of "Critical Infrastructure" policies, but face almost insurmountable problems in trying to move beyond purely symbolic gestures (armed police at airports, Jersey barriers around rail stations, etc.). For here, they face the inevitable fact that, in order to function as infrastructure, today's big technical systems necessarily must be open to a massive flux of use and exchange which cannot be simply controlled, even with the most sophisticated surveillance and information technologies. "Because most mechanisms, structures, and links in world capitalism must be essentially insecure to operate optimally," writes Tim Luke, "defence against the insecurities of all who now live amidst these linked assemblies in big market-driven systems is neither certain nor final."[485] Ultimately, the costs, delays, and reductions in capacity hit at the ultimate bottom line: Big Business, which is now constituted, in many ways, through transnational infrastructural systems for moving raw materials, commodities, capital, information, media and labor power quickly and efficiently across the planet. Totally securitizing infrastructure and its circulations, thus, "adds tremendous cost at the corporate bottom line that few companies are willing to pay."[486]

U.S. STRATEGIC BOMBING: MODERNIZATION'S MIRROR

The mystique of imperial air power depended on relegating the "savage," colonized target population to an "other" time and space, where morale could be decisively shattered by shock tactics and the "moral effect" of aerial bombing. Backward conditions in the colonial laboratory or testing ground guaranteed the technological progress and ongoing modernization of the imperial war machine.[487]

Whilst it has received much less attention than infrastructural terrorism, the targeting of the vital networked systems that sustain modern urban life by states is a great deal more devastating. Such targeting works by "placing a logistical value on targets through their carefully calibrated, strategic position within the infrastructural networks that are the very fibers of modern society." The site-specific bombing of infrastructure, thus, is designed so that effects and impacts "ripple outwards through the network, extending the envelope of destruction in space and time."[488] Such cascading destruction, however, is rarely captured by popular and media commentary, seduced as they are by the contemporary military euphemisms of "precision" targeting and unwanted, accidental, "collateral damage."

The history of the strategic bombing of cities, in particular, can be read at least in part as a history of trying to disrupt their vital systems and infrastructures to bring paralysis to an urbanized adversary. Underpinning the long-standing targeting of civilian infrastructure by aerial bombing—our first and most important example of state infrastructural warfare practices—is a long-standing belief that, in effect, subjecting societies to systematic, aerial bombing is a form of demodernization which is the exact reverse of post-World War II theories of modernization.[489] Just as such theories see "development" as leading societies toward "progress" through successive ages defined by their infrastructure—coal age, electricity age, nuclear age, information age, and so on—so bombing can effectively lead societies, *backwards* as it were, through this linear chain of economic stages by effectively *demodernizing* them.

Thus, just as late twentieth century "development" programs for "developing" nations employed economists and civil engineers, so, bombing programs employ such experts to ensure that destruction brings such imagined reversals into being. One civil engineer advising on U.S. Marine bombing targets during the 2003 invasion of Iraq, for example, noted that "working from satellite photos and other intelligence, supplied pilots with very specific coordinates for the best place to bomb [Iraqi bridges], from a strategically structural point of view."[490]

If one relies on liberal economic theory to herald new technology and infrastructure as deterministically ushering in "progress" and the fruits of new economic ages for whole societies in a simple, linear way, so one is likely to see their systematic devastation as simply reversing such processes, bringing adversaries quickly to their knees. The intimate connections between modernization and development theory on the one hand, and on the other hand demodernization or infrastructural bombing theory, is hammered home forcefully when it becomes clear that, sometimes, the very same experts preside over both. Most notorious here looms the figure of perhaps the most influential U.S. economist of the Cold War: Walt Rostow. On the one hand, he was father of the most important development model of the late twentieth century, outlined in his book *The Stages of Economic Growth*.[491] In this, Rostow suggested a linear and one-way model through which "traditional" societies managed to achieve the "pre-conditions for economic take-off" and then enjoyed the fruits of modernization through the "drive to maturity" and, finally, an "age of mass consumption."

On the other hand, Rostow also played a key part in the U.S. strategic bombing surveys of Japan and Germany and was the most influential National Security Adviser to both the John F. Kennedy and Lyndon B. Johnson administrations from 1961 to 1968.[492] In the latter role, Rostow's incessant lobbying was crucial in gradually extending and increasing the systematic bombing campaigns against civilian infrastructure in the "Rolling Thunder" campaign against North Vietnam. As well as "bombing...countries back through several 'stages of growth,'"[493] this was seen as a means of undermining the communist challenge to U.S. power.[494] Eradicating communism, to Rostow, was necessary because he saw it as a repellant form of modernization. "Communism is best understood as a disease of the transition to modernization," Rostow argued.[495]

This wider notion that bombing, as a form of punitive demodernization, can usher in a simple reversal of conventional models of linear economic and technological progress, is now so widespread as to be a cliché. Curtis LeMay, the force behind the systematic fire-bombing of urban Japan in World War II, famously urged that the U.S. Air Force, which he led at the time, should "bomb North Vietnam back into the Stone Age." He later added that the U.S. "air force ... [has] destroyed every work of man in North Vietnam."[496]

Despite the waning fashionability of the modernization theories to which it provides a dark but much less well known shadow,[497] theories of demodernizing societies through air power remain as popular as ever. Le Mayesque and Rostowian imperatives have come from the lips of many a Strangelove-style U.S. politician, air force leader, or hawkish right-wing commentator since the 1960s. Influential right-wing globalization journalist Thomas Friedman, for example, used a similar argument as NATO cranked up its bombing campaign against Serbia in 1999. Picking up a variety of historic dates that could be the *future* destiny of Serbian society, post bombing, Friedman urged that all of the movements and mobilities sustaining urban life in Serbia should be brought to a grinding halt. "It should be lights out in Belgrade," he said. "Every power grid, water pipe, bridge, road and war-related factory has to be tar-geted.... We will set your country back by pulverizing you. You want 1950? We can do 1950. You want 1389? We can do that, too!"[498] Here the precise reversal of time that the adversary society is to be bombed "back" through is presumably a matter merely of the correct weapon and target selection.

Finally, as U.S. aircraft pummeled Afghanistan in 2002, Donald Rumsfeld famously quipped, with his usual sensitivity, that the U.S. military were "not running out of targets. Afghanistan is."[499] Such a characteristically sick attempt at humor reveals much about the targeting mentality of the U.S. Air Force and the centrality of modern infrastructure as the chosen sites of devastation. Indeed, it shows that *without* a web of modern infrastructures to blast to kingdom come, the U.S. air force literally doesn't know what to destroy at all. One Afghan critic immediately responded, to the parallel claim that the U.S. Air Force would bomb Afghanistan "back into the Stone Age" by suggesting, caustically, that "you can't.... We're already there."[500]

Of course the politics of seeing the bombing of infrastructure as a form of reversed modernization also plays into the long-standing depiction of countries deemed "less developed" along some putatively linear line of modernization as pathologically backward, intrinsically barbarian, unmodern, even savage. "As long as modernization was conceived as a unitary and unidirectional process of economic expansion...," writes Nils Gilman, "backwardness and insurgency would be explicable only in terms of deviance and pathology."[501]

"THE ENEMY AS A SYSTEM"

The central idea shaping the regular U.S. devastation of urban infrastructure systems by aerial bombing in the past two decades has been the notion of the "enemy as a system." Modifying World War II ideas of targeting the "industrial webs" of Germany and Japan to create "strategic paralysis" in war production, this was devised by a leading U.S. Air Force strategist, John Warden,[502] within what he termed his *strategic ring theory*. This systematic view of adversary societies provides the central U.S. strategic theorization that justifies, and sustains, the rapid extension of that nation's infrastructural warfare capability. The theory has explicitly provided the basis for all major U.S. air operations since the late 1990s. The latest U.S. Air Force document on targeting doctrine states that the techniques of infrastructure targeting "often yields useful target sets" and encourages target planners to bomb "infrastructure targets across a whole region or nation (like electrical power or petroleum, oil, and lubricants [POL] production)…, non-infrastructure systems such as financial networks [and] nodes common to more than one system."[503]

Moreover, special weapons have even been developed to destroy civilian infrastructure more effectively. Prime amongst these are so called "soft" or "blackout bombs, which [are] label[ed] the 'finger on the switch' of a country."[504] These rain down thousands of spools of graphite string onto electric transmission and power systems, creating catastrophic short-circuits in the process.[505] As part of the continuous mythology of humanitarianism that pervades military discussions of "precision strikes," these weapons have been widely lauded in the military press as "non-lethal" weapons which create "minimal risk of collateral damage" (i.e., dead civilians)[506] even though their use means that "whole countries, their soldiers and civilian populations can be cast into an isolated darkness within hours of war."[507] As we shall see, though, in the words of Patrick Barriot and Chantal Bismuth, such "high-technology weapons are a long way from proving their innocuousness."[508] Rather, they work to hide the civilian deaths they create by distancing them in time and space from the point of impact. This is convenient

Figure 8.2 John Warden's 1995 Five-Ring Model of the strategic make-up of contemporary societies—a central basis for US military doctrine and strategy to coerce change through air power. Source: Felker (1998) p. 12.

as it tends to remove them from the capricious gaze of the media.

"At the strategic level," writes Warden, "we attain our objectives by causing such changes to one or more parts of the enemy's physical system."[509] This "system" is seen to have five parts or "rings": the leadership or "brain" at the center; organic essentials (food, energy, etc.); infrastructure (vital connections like roads, electricity, telecommunications, water, etc.); the civilian population; and finally, and least important, the military fighting force (see Figure 8.2). Rejecting the direct targeting of enemy civilians, Warden, instead, argues that only indirect attacks on civilians are legitimate. These operate through the targeting of societal infrastructures—a means of bringing intolerable pressures to bear on the nation's political leaders, even though this doctrine contravenes a whole host of the key statutes of International Humanitarian Law.[510]

Air power theorists are clearly well aware that the U.S. policy of destroying civilian infrastructure is likely to lead to major public health crises in highly urbanized societies. In a telling example, Kenneth Rizer, another

U.S. air power strategist, wrote an article in the official U.S. Air Force Journal *Air and Space Power Chronicles*.[511] In it, he seeks to justify the direct destruction of "dual-use" targets (i.e., civilian infrastructures) within U.S. strategy. Rizer argued that, in international law, the legality of attacking dual-use targets "is very much a matter of interpretation."[512]

Rizer writes that the U.S. military applied Warden's ideas in the 1991 air war in Iraq with, he claims, "amazing results." "Despite dropping 88,000 tons [of bombs] in the forty-three-day campaign, only three thousand civilians died directly as a result of the attacks, the lowest number of deaths from a major bombing campaign in the history of warfare."[513] However, he also openly admits that systematic destruction of Iraq's electrical system in 1991 "shut down water purification and sewage treatment plants, resulting in epidemics of gastro-enteritis, cholera, and typhoid, leading to perhaps as many as 100,000 civilian deaths and the doubling [of] infant mortality rates."[514]

Clearly, however, such large numbers of indirect civilian deaths are of little concern to U.S. Air Force strategists. For Rizer openly admits that:

> The US Air Force perspective is that when attacking power sources, transportation networks, and telecommunications systems, distinguishing between the military and civilian aspects of these facilities is virtually impossible. [But] since these targets remain critical military nodes within the second and third ring of Warden's model, they are viewed as legitimate military targets…. The Air Force does not consider the long-term, indirect effects of such attacks when it applies proportionality [ideas] to the expected military gain."[515]

More tellingly still, Rizer goes on to reflect on how U.S. air power is supposed to influence the morale of enemy civilians if they can no longer be carpet-bombed. "How does the Air Force intend to undermine civilian mo-

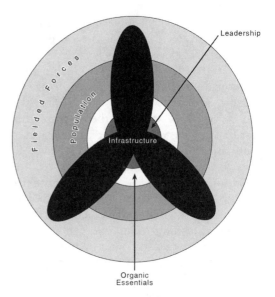

Figure 8.3 'A new model for societal structure'—Edward Felker's adaptation of Warden's Five-Ring Model (see Figure 1), stressing the centrality of infrastructural warfare to post-Cold War US air-power doctrine. Source: Source: Felker (1998) p. 12.

rale without having an intent to injure, kill, or destroy civilian lives?" he asks.

> Perhaps the real answer is that by declaring dual-use targets legitimate military objectives, the Air Force can directly target civilian morale. In sum, so long as the Air Force includes civilian morale as a legitimate military target, it will aggressively maintain a right to attack dual-use targets.[516]

In 1998 Edward Felker, an air power theorist, like both Warden and Rizer, based at the U.S. Air War College, Air University, further developed Walden's model (Figure 8.3).[517] This was based on the experience of the 1991 war with Iraq (code-named Desert Storm) and drew directly on Felker's argument that infrastructure, rather than a separate "ring" of the "enemy as a system," in fact pervaded, and connected, all the others to actually "constitute the society as a whole." "If infrastructure links the subsystems of a society," he wrote, "might it be the most important target?"[518, 519]

"BOMB NOW, DIE LATER"

Such sometimes arcane discussions amongst U.S. bombing planners have massive effects. The U.S. 1991 Desert Storm bombing campaign against Iraq, for example, was targeted heavily against "dual-use" urban infrastructure systems, a strategy that Ruth Blakeley has termed "Bomb Now, Die Later."[520] Moreover, because the reconstruction of life-sustaining infrastructures like electricity, water, and sewerage, was made impossible by the sanctions against the regime that were imposed between 1991 and 2003, it is now clear that the 1991 demodernization of Iraqi metropolitan life—in a profoundly urbanized nation—created one of the largest, engineered public health catastrophes of the late twentieth century. Even U.S. Air Force papers admit that the public health disaster created by bombing Iraq's electricity infrastructure killed at least thirty times as many civilians as did the actual fighting.[521] Along with military and communication networks, urban infrastructures were amongst the key targets receiving the bulk of the bombing. One U.S. air war planner, Lt. Col. David Deptula, passed a message to Iraqi civilians via the world's media as the planes started going in: "Hey, your lights will come back on as soon as you get rid of Saddam!"[522] Another, Brigadier General Buster Glosson, explained that infrastructure was the main target because the U.S. military wanted to "put every household in an autonomous mode and make them feel they were isolated…. We wanted to play with their psyche"[523] As Colin Rowat suggests, for perhaps 110,000 Iraqis, this "playing" was ultimately to prove fatal.[534] Bolkcom and Pike recall the centrality of targeting dual-use infrastructures in the planning of Desert Storm:

> From the beginning of the campaign, Desert Storm decision makers planned to bomb heavily the Iraqi military-related industrial sites and infrastructure, while leaving the most basic economic infrastructures of the country intact. What was not apparent or what was ignored, was that the military and civilian infrastructures were inextricably interwoven.[525]

The prime target of the air assault was Iraq's electricity generating system. During Desert Storm, the allies flew over two hundred sorties against electrical plants. The destruction was devastatingly effective; almost 88 percent of Iraq's installed generation capacity was sufficiently damaged or destroyed by direct attack, or else isolated from the national grid through strikes on associated transformers and switching facilities, to render it unavailable. The remaining 12 percent was probably unusable other than locally due to damage inflicted on transformers and switching yards.[526]

As a result of cascading effects, the destruction of electrical systems meant that Iraq's electrically powered water and sewerage systems collapsed catastrophically. After the war's end, the UN reported that:

> Iraqi rivers are heavily polluted by raw sewage, and water levels are unusually low. All sewage treatment plants have been brought to a virtual standstill by the lack of power supply and the lack of spare parts. Pools of sewage lie in the streets and villages. Health hazards will build in weeks to come.[527]

Because postwar sanctions prevented Iraq from importing the "dual-use" technologies necessary to repair its electricity, sewerage, and water infrastructures, public health crises, and infant mortality rates, spiraled out of control between 1991 and the end of the 1990s. By 1999, drinkable water availability in Iraq had fallen to 50 percent of 1990 levels.[528] As a direct result, UNICEF estimated that, between 1991 and 1998, there were, statistically, over five hundred thousand excess deaths amongst Iraqi children under five—a six-fold increase in death rates for this group occurred between 1990 and 1994.[529] Such figures mean that, "in most parts of the

Islamic world, the sanctions campaign is considered genocidal."[530] The majority of deaths, from preventable, waterborne diseases, were aided by the weakness brought about by widespread malnutrition. The World Health Organization reported in 1996 that:

> the extensive destruction of electricity generating plants, water purification and sewage treatment plants during the six-week 1991 war, and the subsequent delayed or incomplete repair of these facilities, leading to a lack of personal hygiene, have been responsible for an explosive rise in the incidence of enteric infections, such as cholera and typhoid.[531]

SWITCHING OFF THE OCCUPIED TERRITORIES

The second nation-state that has undertaken the most extensive campaigns of systematic infrastructural warfare in recent years has been Israel. These campaigns, however, have had scant attention. Critical discussions of Israeli policies of besieging the Occupied Territories of the West Bank and Gaza have concentrated mostly on civilian deaths caused by air and tank raids; on mass house demolitions and the bulldozing of settlements with massive D9 Caterpillar bulldozers;[532] on the drawing of extremely tight limits on Palestinian enclaves; and the construction of brutal apartheid-style walls, checkpoints, registers, laws, and databases; and on the construction of a parallel world of generously proportioned and expanding Jewish-only settlements linked together with their own private infrastructure and cleared, free-fire "buffer zones."[533]

Much less reported has been a systematic and continuous program by Israeli forces which add a new twist to the geographies and politics of modern siege warfare against urban civilians: the targeting and destruction of modern infrastructure systems. In May 2001, for example, on the eve of "Operation

Defensive Shield," Ben Azri, then Israeli Labour Minister, called for the dismantling of Palestinian roads, utilities, and cultural institutions as a way of "making the Palestinians' life hell."[534] The 2002 Operation Defensive Shield put his words into action. As well as the battles, raids, abductions, and mass demolitions, its central feature—continued in all Israeli operations since—was the deliberate destruction of any symbol of urban modernity of a proto-Palestinian state. Financial damage to infrastructure from the first major offensive alone has been estimated by donors at some $360 million.[535]

During the 2002 operations, water tanks were regularly riddled with bullets. Electronic communications were bombed and jammed. Roads and street furniture were widely bulldozed and destroyed. Computers were smashed; hard discs stolen. Electricity transformers were destroyed. Any cultural or bureaucratic symbol of a proto-Palestinian state was ransacked. Touring the West Bank in April 2002, Amira Hass described the wreckage: "Smashed, burned and broken computer terminals heaped in piles and thrown in yards, server cables cut, hard disks missing, disks and diskettes scattered and broken, printers and scanners broken and missing, laptops gone, telephone exchanges disappeared or vandalised, and paper files burned, torn, scattered or defaced—if not taken." Such destruction, she wrote, "was not a whim, or crazed vengeance. Let's not deceive ourselves—this was not a mission to search and destroy the 'terrorist infrastructure.'"

As Figure 8.4 shows, the main damage to the physical infrastructure of Palestinian cities, roads, water systems, and electricity grids was done using massive 60 ton D9 armored bulldozers. As Mark Zeitoun remarked in August 2002, the huge, behemoth D9s were retrofitted with "special blades and buckets optimized for concrete demolition and a powerful asphalt-ripper in the rear. The resultant power house machinery...is the tool of choice for destroying electrical grids, digging up buried water and sewerage services, taking out shop fronts and demolishing cars"[536] (Figure 8.4).

Like the U.S. bombing of Iraq, such actions are a direct reflection of changes in Israeli military doctrine where the systematic targeting of civilian infrastructure is seen as a means of coercing adversaries in the "non-traditional" wars against insurgents and supportive civilian populations in cities. Israeli doctrine here has clearly been heavily influenced by U.S. theories surrounding John Warden's work, already discussed, where adversary societies are seen as "systems of systems" and the targeting of urban infrastructure is one means of launching "effects based operations" to psychologically coerce entire populations. Israeli military theorists now talk in particular of targeting infrastructure as the key means of undertaking "diffused warfare" where there are no obvious front lines. "Rather than being defined by parameters of front lines and home fronts," retired Israeli Admiral Yedidia Groll-Yaari and Haim Assa, wrote in a recent paper on Israeli "geo-strategy," "the nature of future conflicts for nation states will be determined by legitimate objectives and desired effects in a *multitude of contact points*—be they military or civilian, infrastructure."[537] Thus, as in U.S. doctrine, destroying civilian infrastructure is seen by Israeli military planners as one of the few ways of bringing pressure to bear on the actions of stealthy and usually hidden insurgents and fighters.

Strangling Gaza

It could, rightfully, be a cause of shame to the world. But the world, besieged by violence and injustice, hardly notices it.[538]

Whilst Lebanon's urban infrastructure was devastated by Israel in 2006, as part of this new strategy of launching "diffuse war" against Hezbollah, it is the Gaza Strip that

Figure 8.4 An 60-ton Israeli D-9 bulldozer destroying Palestinian road and water infrastructures during the 'Operation Defensive Shield' invasion of the West Bank, April 2002. Source: A Palestinian activist who wishes to remain anonymous.

provides perhaps the most startling example of the effects of this new Israeli doctrine, developed since 2006.[539] For it is within Gaza space that Israel has pushed to extremes its "diffuse warfare" strategy for the Occupied Territories. This combines physical, hermetic closure,[540] the prevention of circulation, intensive aerial surveillance, continuous air raids, the devastation of modern infrastructure, and devastating incursions by squadrons of tanks backed up by artillery assaults.

The Israeli military's idea behind "diffuse warfare" is to combine withdrawal from Gaza and the maximization of at-a-distance military control with a complete absolution of political, legal, social, or moral responsibility for the fates of Gaza's 1.4 million inhabitants.[541] "The Gaza Strip," writes Darryl Li, is thus a "space where Israel tests and refines various techniques of management, continuously experimenting in search of an optimal balance between *maximum control* over the territory and *minimum responsibility* for its non-Jewish population."[542] The strip provides a kind of dystopian proving ground for practices that could become increasingly relevant in the West Bank as Palestinian life becomes increasingly fragmented into an archipelago of isolated "Gaza Strips."[543]

Israel's new strategy has transformed what was, in effect, a giant open-air prison into a massive, besieged city-strip where there are no apparent prospects for the siege—which is demographically inspired—to be lifted. The strangulation of Gaza has dramatically intensified following the evacuation of Jewish settlements from the Strip in late summer 2005, and the election of a Hamas government there in January 2006.[544] The prime cause of the present crisis was that "the Palestinian people went to the polls, participated in free, fair, and transparent democratic elections unmatched in the Arab world, but voted for the *wrong party*"[545]—Hamas. Israel, the European Union, the United States, and other aid donors then decided to apply economic, tax, and aid sanctions against what

was immediately cast as a "terrorist state." Israel also declared that Gaza was from there on in a "hostile territory" with which it was "at war."[546]

Two days after the taking prisoner of Corporal Shalit by Palestinian fighters in Rafah on June 25, 2006, Israel launched its "Summer Rain" offensive against Gaza. At the outset of the attacks, Israeli Prime Minister, Ehud Olmert, claimed that the operation was not aimed "to mete out punishment but rather to apply pressure so that the abducted soldier will be freed."[547] During the Summer Rain attacks, the Israeli Defense Force also claimed that the goals of the operations were "to stop the terrorist organizations that relentlessly fire [home-made Kassam] rockets" over the border into Israel, and that the operations were "tailored to avoid civilian casualties."

The latter seems a particularly ridiculous claim: the use of artillery and bombing in one of the world's most densely populated urban environments will *necessarily* maim and kill large numbers of civilians. This result is not accidental or "collateral" in that those who take these actions know this inevitability full well. Between June 28 and September 13, 2006, 290 people, mostly civilians, were killed in Gaza by Israeli actions; 135 of these were children.[548] In addition, 750 people sustained injuries leaving them permanently disabled. Later Israeli incursions in 2008, ostensibly designed to stop the firing of home-made rockets, killed another 323 Palestinians in response to the deaths of seven Israelis, only two of whom were civilians.[549] In the seven years between 2001 and 2008 there were seven civilian deaths in Israel caused by the home-made rockets.[550]

Despite the press releases of the Israelis and their military, it is hard not to conclude that these operations were designed as a massive exercise in collectively punishing Gazans. Revealingly, these press releases stressed that "it must be recalled that the Palestinian people themselves elected a government led by Hamas, a murderous terrorist organization."

Some of Israel's policies, such as the deliberate creation of sonic booms low over Gaza by aircraft at night—which have particularly traumatic effects on children—were clearly acts to deliberately terrorize the population. However, the tightening of the *infrastructural* siege and destruction were far more devastating. Food imports were curtailed (a devastating gesture in a city relying heavily on food imports and food aid for survival).[551] Fuel and power supplies were also cut off. Bridges and roads were devastated by bombing. All six transformers of Gaza's main electricity generating station, which provided half of Gaza's power, were also bombed, which had knock-on effects in the dwindling of pumped water and sewerage services (see Figure 8.5).[552] "Even before the ground invasion" of the Rafah area, "the Israeli air force (IAF) bombed a power station, which disrupted the electricity supply for large sections of the Gaza Strip, as well as the water supply." As

in Iraq in the period after 1991, deelectrification meant that water and sewerage systems were disabled, hospitals could not function normally, businesses were often completely disabled, and regular urban life ceased to be possible.[553] Finally, parts for crucial repairs to devastated infrastructures were subject to sanctions.[554]

In the aftermath of the attacks, Karen AbuZayd, United Nations Relief Works Agency (UNRWA) Commissioner-General in Gaza, argued that the Strip was "on the threshold of becoming the first territory to be intentionally reduced to a state of abject destitution, with the knowledge, acquiescence and—some would say—encouragement of the international community."[555] She added that "the decision to limit fuel and potentially electricity to the general population constitutes a form of collective punishment which directly contravenes international humanitarian law."[556] Imogen Kimber of

Figure 8.5 Destroyed electricity transformer in the Gaza Strip following Israeli bombings, 2006/.

the International Middle East News Center meanwhile called the whole attack a "sick madness" in which "power cuts caused by the deliberate bombing of the electricity supply by the Israeli air force leave doctors with a near impossible task in attempting to treat the injuries and ill health."[557]

As a result of "Summer Rain," many of Gaza's health facilities ceased to function, as they had no generators (which were often useless in any case because of fuel shortages). The public health results were fairly immediate. The World Health Organisation (WHO) reported that "the total number of cases of watery and bloody diarrhoea amongst refugees for the last week in June and the first week in July [2006] ha[d] increased by 163 percent and 140 percent compared to the same period last year."[558] Rates of anemia amongst children also skyrocketed.[559] Already high rates of malnutrition and stunted growth amongst Gazan children were increased further.[560]

Within a few months, the public health system of Gaza was on the verge of complete meltdown. Extremely ill patients were no longer able to travel to Israel for care. Dialysis patients started dying because dialysis sessions had to be reduced in frequency because of lack of medicine and electricity blackouts caused by the bombing of generating stations.[561] At the end of March 2007, in perhaps the ultimate symbol of the strangulation of Gaza, the backup of raw sewerage was so bad that some neighborhoods were flooded as retaining banks collapsed. In the North of the Strip, five people drowned in one sewerage flood. Gazans, it seems, were reduced to a people literally drowning in their own shit.[562]

"Steel Rain," however, was a small-scale operation compared to the "Cast Lead" campaign launched by the Israeli military in December 2008 to January 2009. This invasion massively scaled up the level of devastation caused by regular tank, artillery, and aerial bombing raids. It resulted in at least 1,500

dead, over 5,000 severely injured, 21,000 apartments bulldozed or destroyed, and virtually all of Gaza's modern infrastructure in a state of complete destruction.[563]

STATE CYBERWARFARE

According to William Church, ex-Director of the now-defunct Center for Infrastructural Warfare Studies, the next frontier of state infrastructural warfare will involve the development of capacities to undertake coordinated "cyberwarfare" attacks.[564] "The challenge here," he writes, "is to break into the computer systems that control a country's infrastructure, with the result that the civilian infrastructure of a nation would be held hostage."[565] Church argues that NATO considered such tactics in 1999 in Kosovo and that the idea of cutting Yugoslavia's Internet connections was raised at NATO planning meetings, but that these options were rejected as problematic. But within the U.S. emerging doctrine of "Integrated Information Operations" and infrastructural warfare—which encompass everything from destroying electric plants, dropping Electronic Magnetic Pulse (EMP) bombs that destroy all electrical equipment within a wide area, developing globe-spanning surveillance systems like Echelon, to dropping leaflets and disabling Web sites—a dedicated capacity to use software systems to attack an opponent's critical infrastructures is now under rapid development.

Deliberately manipulating computer systems to disable opponents' civilian infrastructure is being labeled Computer Network Attack (CNA) by the U.S. military. It is being widely seen as a powerful new weapon, an element of the United States' wider "Full Spectrum Dominance" strategy.[566] Whilst the precise details of this emerging capability remain classified, some elements are becoming clear.

First, it is apparent that a major research and development program is underway at the

Joint Warfare Analysis Center at Dahlgren (Virginia) into the precise computational and software systems that sustain the critical infrastructures of real or potential adversary nations. Major General Bruce Wright, Deputy Director of Information Operations at the Center, revealed in 2002 that "a team at the Center can tell you not just how a power plant or rail system [within an adversary's country] is built, but what exactly is involved in keeping that system up and making that system efficient."[567]

Second, it is clear that, during the 2003 invasion of Iraq, unspecified offensive computer network attacks were undertaken by U.S. forces.[568] Richard Myers, Commander in Chief of U.S. Space Command, the body tasked with Computer Networked Attack, admitted in January 2000 that "the U.S. has already undertaken computer networked attacks on a case-by-case basis."[569] A National Presidential Directive on Computer Network Attack (number 16), signed by George Bush in July 2003, demonstrated the shift from blue-sky research to bedded-down doctrine in this area.

In 2007 it was also announced that the U.S. Air Force had established a new "Cyber Command," located at 8th Air Force at Barksdale Air Force Base, Louisiana. This was tasked with both "cybernetwork defense" for the U.S. Homeland and "cyberstrike" against adversary societies (another term for computer network attack).[570] The purported aim of the five year program, which effectively attempts to militarize the world's global electronic infrastructures, is to "gain access to, and control over, any and all networked computers, anywhere on Earth."[571] Lani Cass, previously the head of the Air Force's Cyberspace Task Force, and now a special assistant to the Air Force Chief of Staff, reveals that these latest doctrines of cyberattack are seen to be a mere continuation of the history of striking societal infrastructure with air power. "If you're defending in cyber [space]," he writes, "you're already too late. Cyber delivers on

the original promise of air power. If you don't dominate in cyber, you cannot dominate in other domains."[572]

These efforts to bolster the U.S. cyberwar capabilities were significantly bolstered by a series of massive "denial of service attacks" against Estonia in Spring 2007.[573] In apparent revenge for the shifting of a statue dedicated to the Soviet war dead in Talinn, these crippled the Web sites of Estonia's prime minister as well as Estonian banks. They seemed to emanate, at least in part, from hackers linked to the Russian State. U.S. strategic planners are also closely monitoring the growing ability of China's armed forces to launch sophisticated and continuous cyberwarfare attacks as part of Chinese doctrine of "unrestricted" or "asymmetric" warfare.[574] Such risks led NATO leaders to say, in 2008, that they took the risks of cyberwarfare attacks as seriously as a missile strike.[575]

The concern of U.S. military planners is that the proliferation of cyberwarfare might open up advanced and high-tech economies, which rely on the most interdependent, dense, and computerized infrastructure systems, to attack from a wide range of state and non-state organizations operating at a diversity of scales. If state or terrorist cyberattacks were to become common, Steven Metz, of the U.S. Strategic Studies Institute argued, "the traditional advantage large and rich states hold in armed conflict might erode. Cyberattacks require much-less-expensive equipment than traditional ones. The necessary skills can be directly extrapolated from the civilian world.... If it becomes possible to wage war using a handful of computers with internet connections, a vast array of organizations may choose to join the fray."[576] Metz even suggests that such transformations could lead nonstate groups to attain power equivalent to nation-states, commercial organizations to wage cyberattacks on each other, and cyber "gang wars" to be played out on servers and network backbones around the world rather than in ghetto alleys."[577]

In an interconnected world, however, where infrastructures systems tightly connect both with each other and with other geographical areas, it is strikingly clear that the effects of such cyberwarfare attacks are likely to be profoundly unpredictable. Whilst attacking Iraq in 2003, for example, it is now clear that U.S. Air Force computer network attack staff considered the complete disablement of Iraqi financial systems using computer network attack techniques. But they apparently rejected the idea because the Iraqi banking network was so closely linked to the financial systems network in France. This meant, for example, that an attack on Iraq might easily have led to the collapse of the Europe's ATM machines.[578] "We don't have many friends in Paris right now," quipped one U.S. intelligence officer on the emergence of this decision. "There is no need to make more trouble if [then French President] Chirac won't be able to get any Euros out of his ATM!"[579]

"WAR IN THIS WEIRDLY PERVIOUS WORLD"

How civilising can war indeed be, if it kills thousands of people and destroys the very infrastructure of civilisation?[580]

This chapter has sought to assert the central place of efforts to disrupt and destroy urban infrastructure within contemporary spaces of war and terror in the context of what computer scientist Phil Agre has called our "weirdly pervious" world. Whilst long neglected and taken for granted—as long as they work—the everyday infrastructures of urban life are increasingly at the heart of contemporary political violence and military doctrine. In our rapidly urbanizing world, they are the main targets for catastrophic terror attacks. They are increasingly central in the doctrines of advanced Western and non-Western militaries; they lie at the heart of

contemporary means for projecting state terror. Whilst we have focused in detail here on the deadly cascading effects on Iraq's urban civilians of a brief frenzy of electrical devastation in 1991, on Israel's attacks on Gaza and the Occupied Territories since 2002, and on nascent powers of state cyberwarfare, we might also have focused on Russia's regular threats to turn off the energy supplied to entire nations as a source of its resurgent political power, or on China's development of new approaches to state warfare based on nonmilitary efforts to manipulate societal infrastructures. We might also stress that there is also increasing evidence that nation-states are already actively engaging in low-level computer network attacks on a more or less continuous basis, a process that blurs the boundary separating warfare and economic competition.

We have seen in this chapter that the realities and effects of efforts by state militaries to supposedly minimize "collateral" damage through targeting everyday civilian infrastructure are far from the ethereal and abstract niceties portrayed in military doctrine and theory. Rather, the experiences of Iraq, Gaza, and Lebanon remind us forcefully that the euphemisms of such theory distract attention from how in highly urbanized and modern societies, the targeting of essential infrastructure—especially infrastructure and water—is a means of killing the weak, the old, and the ill just as surely as carpet bombing them. The difference, of course, is that the deaths are displaced in time and space from the capricious gaze of the mainstream media, who still are often seduced by the geek-speak of military press conferences about "effects based operations," minimizing "collateral damage," targeting "terrorist infrastructure," or putting "psychological pressures" on adversary regimes.

Three key points arise from this discussion. The first is that strategies of infrastructure disruption through political violence require us to reconsider prevailing notions of war. Contemporary intersections of political violence and infrastructure further blur traditional binaries separating "war and peace," the "local" and the "global," the "civil" and "military" spheres, and the "inside" and "outside" of nation-states. Such trends suggest that potentially boundless and continuous landscapes of conflict, risk, and unpredictable attack are currently emerging, as the everyday infrastructures sustaining urban life on an urbanizing planet, usually taken for granted and ignored, become key mechanisms for the projection of state (and nonstate) violence. Many contemporary military theories and doctrines, as we have seen throughout this book, conclude that "war, in this sense," writes Phil Agre, "is everywhere and everything. It is large and small. It has no boundaries in time and space. Life itself is war."[582]

The problem with such strategies, of course, is that they implicitly push for a deepening militarization of all aspects of contemporary urban societies. Questions of security "come home" with a vengeance: They penetrate utterly the everyday, the practices, architectures, and politics of urban life. In the process, they become less and less concerned with territorial state vs. state conflicts during formalized wars and battles and focus instead "on the civic, urban, domestic and personal realms" within boundless conceptions of unending risk.[583] War, in this broadest sense, suggests Phil Agre becomes a continuous, distanciated event, without geographical limits, that is relayed live, 24/7, on TV and the Net.[584]

Our second key point is that there is a very real risk here that the "security" of urban societies against new notions of permanent and boundless infrastructural war becomes such an overpowering obsession that it is used to legitimize a reengineering of the everyday systems that are purportedly now so exposed to the endless, sourceless, boundless threat. The worry here is that constructions of endless, boundless threats and limitless, boundless war provide the legitimation for the whole-

sale hollowing out, or potentially even eradication, of democratic societies. "The military intellectuals' new concept of war is flawed because it starts from the military and simply follows the logic of interconnection until the military domain encloses everything else."[585] The danger of this scenario is that *everything* is viewed as an element of warfare. There is, indeed, nothing *outside of* these boundless and infinite conceptions of war. The acceptance of such a view provides ripe conditions for profoundly antidemocratic shifts toward soft, incipient, or hard fascism, as far-right political coalitions demand the suspension of due process, legal norms, and democratic rights and processes by scapegoating a wide range of immanent and ubiquitous threats which lurk invisibly within the technical interstices of urban life.

A dominant characteristic of the "war on terror," certainly, has been the endless portrayal of the everyday sites, spaces, and systems of the city as domains where lurking Others might jump out at any time offering existential threats to cities and civilizations "from within."[586] In portraying the risks of terrorism as both acts of "war" and existential societal threats, rather than as international crimes with massive public safety risks, it is easy, of course, to justify global and unending war, extending imperialism, racialized state violence, preemptive incarceration, authoritarian legislation, and the radical suspensions of legal and juridical norms. All of these are consistent with recent trends in societies like the United States and Britain toward what many critical commentators deem "soft" or "incipient" fascism.[587]

The danger here, of course, is that the chipping away of democratic rights, and the progressive expansion of national security states toward globe-spanning surveillance operations which attempt to parallel the infrastructural mobilities of the world, becomes as endless as the purported threat. This is driven by the construction of a continuous series of real or chimeric infrastructural terror threats, fanned by the flames of sensationalist, voyeuristic, and jingoistic media. To Phil Agre:

war in the new sense—war with no beginning or end, no front or rear, and no distinction between military and civilian—is incompatible with democracy…. The conditions of war [become] almost identical with the social vision of conservatism…. The new military doctrine of war as total phenomenon, war without boundaries, is nothing except conservatism…. Instead of permanent, total war, conducted under rules that subordinate democracy to an authority that draws its legitimacy from an absolute evil or foe. We need a conception of permanent, total security [based on a sustained re-design of everyday infrastructure], conducted under rules that keep the ends squarely in view.[588]

Agre argues that the use of discussions about infrastructure and war to sustain limitless and boundless militarization is a massive and urgent threat to democracy itself. "We can't get a new concept of war [in a context of infrastructural warfare] without getting a new concept of democracy."[589]

Ultimately, then, ideas of "security" need to be radically rethought so that the human security of subjects within cities, infrastructural systems, biospheres, and social worlds is the central object of governance,[590] not some "hard" notion of "national security based on permanent war and hypermiltarization, the retreat into militarized enclaves, and the application of military paradigms to all walks of life and governance." Crucially, Agre argues that "the important thing is to draw a distinction between military action, as the exercise within a framework of international law of the power of a legitimate democratic state, and war, as the imposition of a total social order that is the antithesis of democracy, and that, in the current technological conditions of war, has no end in sight."[591]

Our final key point is that efforts by state militaries to systematically destroy the essential infrastructures of those urban societies

deemed adversaries require major discursive work, as well as the effort to, as Captain John Bellflower of the U.S. Army has put it, "put steel on the target."[592] Most often, of course, the targeting of the basic water, sewerage, electricity, transport, and communications infrastructures sustaining every aspect of modern urban life in an entire society is targeted as a putative means of "destroying the infrastructure of terrorism." Both Israel and the United States have long legitimized their systematic demodernization of whole societies in this way and Palestine, Lebanon, and Iraq amongst others have suffered the consequences. These, as we have seen, spiral way beyond the moment and location of the attack and cause widespread death, disease, poverty, and economic collapse.

The tragic and perverse irony here is that terrorist attacks against Western or Israeli cities actually rely not on the basic modern services in Iraq, Palestine, or the Lebanon to launch their attacks, but on Western bus systems, airline networks, tube and subway trains, mobile phone infrastructures, and so on. The process through which rich countries launch "war" against "terrorist infrastructure" within poor countries thus does little but radicalize and immiserate entire generations, dramatically adding to the poor and willing recruits ready to launch more terrorist attacks against the West:

The phrases "defeating terrorist states" and "destroying the infrastructures of terrorism" turn out to mean, simply, "defeating states" and "destroying infrastructure." If anything, reducing the functioning society to anarchy by destroying its infrastructure and killing great numbers of its citizens is likely to increase whatever legacy of grudge and grievance is already in place."[593]

The systematic demodernization of whole societies in the name of "fighting terror," in fact, involved a deeply ironic self-fulfilling prophecy. As the geographer Derek Gregory[594] has argued, drawing on the work of Italian philosopher, Georgio Agamben,[595] the demodernization of entire Middle Eastern cities and societies, in both the Israeli wars against Lebanon and the Palestinians, and the U.S. "war on terror," are both fueled by similar, Orientalist discourses. These are based on binary imaginations of geography where "our" space—a human and valorized space of the homeland city—is completely separable from "their space"—reduced to what Derek Gregory has called "a space empty of people" that can be targeted from afar, or from the air, without having to confront the civilian horrors on the ground that result.[586] Such logics of targeting revivify long-standing Orientalist tropes and work by "casting out" ordinary civilians and their cities—whether they be Kabul, Baghdad, Gaza, or Nablus—"so that they are placed beyond the privileges and protections of the law so that their lives (and deaths) [are] rendered of no account."[597] Here, then, beyond the increasingly fortified homeland, "sovereignty works by *abandoning* subjects, reducing them to bare life."[598]

In then forcibly creating a kind of chaotic urban hell in cities that are systematically demodernized and "switched off," perversely, this state violence produces what Orientalist representations depict: a chaotic and disconnected urban world "outside of the modern, figuratively as well as physically."[599] If Western culture has long been predicated on the presence of the antimodern figure of the Oriental, it becomes very easy to model Western warfare as the means to systematically demodernize Orientalized cities and societies—all in the name of defending the infrastructures of the putative "homelands" from attack. The result is another self-perpetuating cycle because, not surprisingly, the anger and despair of those living in switched off cities quickly turns, and can be readily exploited, to radicalization, a willingness to launch terroristic violence against the perpetrators of their plight. And so, of course, the cycle goes on.

CHAPTER 9

Infrastructure, Interruption, and Inequality: Urban Life in the Global South

Colin McFarlane

INTRODUCTION

In much debate on cities in the Global South, infrastructure is synonymous with breakdown, failure, interruption, and improvisation. The categorization of poorer cities through a lens of developmentalism has often meant that they are constructed as "problem."[599] These are cities, as Anjaria has argued, discursively exemplified by their crowds, their dilapidated buildings, and their "slums."[600] A sense of a burgeoning population has led to scholastic and public accounts of these cities as, in Seabrook's phrasing, uncontainable and inadequate: "The terms in which the cities are discussed—urban 'explosion', 'catastrophe'—tend to assimilate them to natural disasters; they are problems crying out first for relief, and then for solutions."[602] This is a tendency echoed in some recent critical accounts of informality as a key part of the urban condition. For instance, writing about Davis's popular *Planet of Slums*, Agnotti has questioned its "apocalyptic rhetoric" and Dickensian imagery that writes out the agency and complexity of the cities in question: "In Davis' dismal descriptions of urban poverty, there are no people or social forces capable of challenging the social order."[602]

As Seabrook points out, against this backdrop of despair, there is an alternative view that conceives "slum" as inventive and en-trepreneurial, a space of courage and endurance.[603] For Davis, who distances himself from debates emphasizing the empowered or entrepreneurial subject, "slums" produce quite different kinds of social agency. In these settlements—in his view the fastest growing and most unprecedented social class on earth—the "slum dweller" is caught amidst the increasing grip of the Pentecostal Church (especially in South America and parts of Africa) and militant Islam (especially in the Middle East, South Asia, and North Africa). Amid this mood of hopelessness, he asks: "To what extent does an informal proletariat possess that most potent of Marxist talismans: 'historical agency'?"[604]

There is a latent danger in these debates of environmental determinism, and in particular of equating infrastructure failure with social or political failure (Anjaria, 2008). In this sense at least, Davis's text inadvertently enjoins a long history of contextualizing infrastructure with societal development, from modernization theory's preoccupations with "stages" of technological development (e.g., Rostow, 1960), to pervasive accounts of city "success" in economic and developmental explanations such as those of the World Bank.[605] This notwithstanding, Davis raises a key set of issues around the relationship between infrastructure interruption and collective culture. Writing on Lagos, Gandy considers the potential role of infrastructural networks in forging social collectivities through the "binding of space" in contexts of poverty, social fragmentation, and governmental failure. For Amin, the quality of urban infrastructure cannot but affect urban civic culture: "When the basics of shelter, sanitation, sustenance, water, communication and the like are missing, the experience of the city, of the commons and of others, is severely compromised, producing solidarities of largely an exclusionary and wretched nature."[606] The relationship between agency, settlement, and infrastructure interruption raises important questions about how different actors variously cope with, and seek to address, infrastructure lack or inadequacy, from the World Bank's invocation of infrastructure privatization, to the Indian state's focus on more elite infrastructures like air, rail, and flyovers over, for example, sanitation and public transport, and the politics of urban movements like Slum/Shack Dwellers International.[607]

This chapter argues that infrastructure interruption reflects and reproduces urban inequality. It does so by highlighting three key relations between inequality, interruption, and infrastructure. First, in the case of largely unforeseen infrastructure crises, it is often in the *responses* of local governments and infrastructure managers that we can analyze the reinforcement or transformation of power relations. Second, crises are themselves mediated by and often exacerbate existing forms of inequality. Third, and quite apart from more typically everyday forms of interruption such as power outages, particular forms of "crisis" are becoming increasingly common, raising the question of whether they actually constitute a "new normality" for some groups.[608] The first part of the chapter sets the scene by exploring the nature of existing infrastructure connection and interruption in cities in the South, drawing attention in particular to how infrastructure becomes understood in relation to circulation and progress, and highlighting the growing importance of

processes of improvisation and privatization as responses to infrastructure interruption. The chapter goes on to consider the three links between infrastructure and inequality highlighted above before exploring them in detail through the case of the 2005 monsoon flood crisis in Mumbai. This last part of the chapter draws on research conducted in Mumbai following the monsoon throughout 2006. In contrast to the image of social collapse in the face of infrastructure failure, the Mumbai monsoon revealed, as Anjaria has argued, not social despair or empowered entrepreneurism, but urban generosity and hospitality in the face of infrastructure crisis.[609] But as an infrastructure crisis, the monsoon resulted in a variety of responses beyond generosity, including depictions of "slums" as malaise in need of removal, and wide-ranging debates about the possibilities and limits of infrastructure, urban development, and the city itself.

INFRASTRUCTURE CONNECTION: CIRCULATION AND IMPROVISATION

There are, of course, many different forms and ways of conceiving "interruption" that vary within and between cities. There are large-scale urban infrastructure crisis such as those caused by flooding, landslides, earthquakes, or war, and more everyday forms of interruption that have received relatively little attention in urban studies, including power outages, water contamination, fire, collapsed buildings, traffic congestion and road blocks, and construction and demolition.[610] The causes are various, from pollution from industrial plants and lack of maintenance or repair, to state neglect, lack of legal tenure or enforced regulatory frameworks, to state neglect and processes like weathering. The roles and limits of the state are paramount here, as Davis has argued: "'Fragility' is simply a synonym for systematic government neglect of environmental safety."[611] Crucially, interruption is socially mediated, so that, for example, it is very unlikely that people in the poorest of Mumbai's informal settlements will have a similar conception and experience of interruption of water or electricity, than the middle classes living and working in parts of the city's relatively wealthy southern neighborhoods. The specific nature and experience of infrastructure interruption varies a great deal in type, impact, and response, depends on the nature of existing infrastructure connections across the city, and reflects patterns of urban inequality. Consider, for example, these two quotes from newspaper reports of public transport in Delhi, the first detailing a public bus journey, the second an underground train journey on the new air-conditioned underground metro:[612]

> Right from managing a foothold to getting down after a 45-minute journey, the entire episode was all a matter of survival…the stench from the roadside drain filled the bus…the journey was painfully slow due to red lights at every crossing.[613]

> Imagine travelling in the middle of North Delhi at Rs 8 (about US $.20) and covering the distance between Trinagar and Shahdara in 19 minutes, instead of a bumpy, push-and-shove ride in a rickety bus that will take anywhere between 40 minutes and an hour.

An average journey on the metro is far more expensive than a comparable journey by bus, in some cases by half as much extra. The metro was built beneath the debris of informal settlement demolition along the Yamuna River. But the public symbolism of the metro has been largely positive, partly because it is characterized by *non*interruption, and is opposed to the "bumpy, push-and-shove" cheaper public buses above ground. This circulatory capacity resonates beyond simply comfort or reliability—it has historically infused ideals about the city, modernity, and progress.[614] Siemiatycki argues: "Beyond the movement of people, the metro has been

officially sanctioned as a vehicle for inculcating a culture of discipline, order, routine and cleanliness in Delhi. These ideals are in sharp contrast to the congested, unpredictable, chaotic pace of life in Delhi today."[615] The Delhi Metro is perceived in contrast to the sprawling, haphazard, unclean infrastructure above ground, and in this sense the *spectacle* of infrastructure free from interruption matters a great deal:

> Inside the metro stations, aesthetics that include an open concept layout, technologically advanced no-touch turnstiles, security cameras and well-appointed station trimmings project an image of progress, order, cleanliness and security. The silver trains with their sleek industrial design, automatic doors, digital signs and climate control are a tangible embodiment of the future. While life outside the station walls may be uncivil, aggressive and crowded, inside behaviour is calm, unthreatening and comfortable. This is the vision for a modern Delhi, as promoted by the politicians and technocrats leading the development of the metro in their media sound bites and inauguration speeches.[616]

Not only is the dated and unreliable infrastructure above ground discursively linked to the "uncivil, aggressive and crowded," the city government has been known to point to the metro as evidence of Delhi's entry into the realm of "world-class cities": modern, global, and reliably on the move. At stake in debates and processes of interruption and circulation are notions of the city as modern and equal. In this context, the focus of infrastructure spending is a key issue. In Indian cities more generally, a great deal of current infrastructure spending is focused on main roads, airports, metros, and cellular connections. There were 7.8 million new cellular connections across India in September 2007 alone.[617] Much of this investment is designed to attract investment to India or retain it. The Indian government, and the World Bank and IMF, is acutely aware that, for example, over one third of Indian firms surveyed in the 2004 Infrastructure Climate Assessment cited infrastructure as a "major" or "severe" obstacle to business expansion.[618]

If continuous efforts of infrastructural improvisation and repair characterize the life of many urban infrastructures,[619] especially in the Global South, breakdown is nonetheless inconceivable in the public imagination of many networked infrastructure systems. For example, when the Hong Kong underground, one of the most reliable train systems in the world, was disrupted in May 1996 by a track-circuit defect, the response was one of public and governmental shock.[620] The expectation of uninterrupted infrastructure is particularly associated with elite expectations of the modern, circulatory city. Swyngedouw writes of how the metaphor of circulation and infrastructure has itself historically circulated amongst different public and governmental domains: "New principles of city planning and policing were emerging based upon the medical metaphors of 'circulation' and 'flow'."[621]

If we have witnessed in recent years a proliferation of elite urban enclaves in cities in the Global South, these enclaves are often discursively portrayed as spaces of noninterruption, as located physically within or on the edge of the city, but outside the infrastructural unreliability of the rest of the city (Cowen, this volume). These elite enclave developments are informed by a global policy comparativism driven by consultancy and often intense competition for capital. New elite towns are emerging that imitate other forms of urban enclave, explicitly aping rival global locations like Dubai's Jebel Ali, Malaysia's Bandar Nusajaya Industrial Park, and the Special Economic Zone phenomenon in China. If gated enclaves are the signature form of this exclusive urbanism, they often entail or reflect particular biopolitical imaginaries and choices that mean their significance extends beyond simply inflated real estate.

As Bunnell and Coe argue in relation to Cyberjaya, part of Malaysia's elite Multimedia Super Corridor development:

> We suggest that buying into the cybercorridor can in part be understood in terms of participating in modes of (self-)government based on lifestyle choice through practices of consumption. "Intelligent investment" (as one advertisement put it) in a "dot.com property" in Cyberjaya is thus not just about buying real estate—though it definitely is partly that—but is also about investing in oneself and one's family for a supposedly immanent information age.[622]

Key to this "information age" is the imaginary of a circulatory, uninterrupted network infrastructure, from roads and electricity to Internet and air conditioning. Another example here is the Special Economic Zone, Magarpatta City, located between Mumbai and Pune, which has been built and advertised as a gated elite enclave that constitutes, in the city's Web site phrasing, "a new way of life for the networked society of the new millennium" (www.magarpattacity.com) (see Figure 9.1). Amongst other provisions, Magarpatta City boasts a "cybercity," a "garden city," and a "round the clock centralised security system," and has consistently emphasized the reliability of its electricity, Internet, and road infrastructures.

Indeed, written across Mumbai's skyline is a litany of billboards advertising future urban escapisms, and the promise of high-end, circulatory, and uninterrupted infrastructure networks, as Figure 9.2 shows.

Beyond these stark divisions in urban infrastructural imaginaries and practice, there is a high degree of variability in the experience of interruption within and between cities in the Global South. To take a different example, one comparative study of electricity in Nigeria, Thailand, and Indonesia found that 92 percent of Nigerian firms had their own generators to supplement the inadequate public supply, while the figure was 66 percent in Indonesia and only 6 percent

:: MAGARPATTA CITY PLAN ::

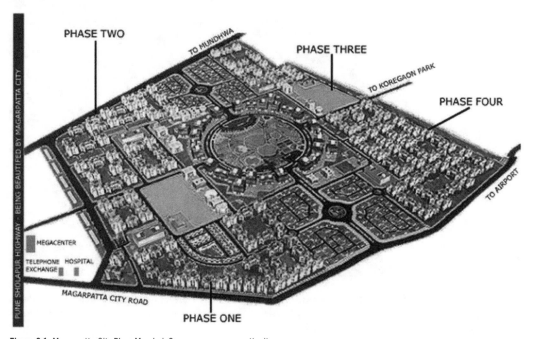

PHASE TWO

PHASE THREE

PHASE FOUR

TO MUNDHWA

TO KOREGAON PARK

TO AIRPORT

PUNE SHOLAPUR HIGHWAY - BEING BEAUTIFED BY MAGARPATTA CITY

MEGACENTER

TELEPHONE HOSPITAL
EXCHANGE

MAGARPATTA CITY ROAD

PHASE ONE

Figure 9.1 Magarpatta City Plan, Mumbai. Source: www.magarpattacity.com

in Thailand, although the quality of electric power in Thailand was similar to that of Indonesia.[623] While the impacts of these costs on manufacturing are felt often profoundly by smaller firms, larger companies can often absorb the costs with relative ease. Interruption is mediated by inequality, and poorer groups often have the least capacity to cope with interruption and to improvise temporary or long-term solutions. Mumbai, for instance, has a huge thirst for electrical power. The current supply for the city is 2,200MW, but throughout June and the following months, as the demand on air conditioners in particular escalates, demand is between 2,900MW and 3,100MW.[624] This leaves suppliers in the position where they must either opt for load sharing, as many smaller Indian towns do, or buy in power from outside. With other large Indian cities in similar positions, buying-in creates a lucrative market for power utilities, the costs

of which are typically passed on to consumers and felt especially by the poor.

In contexts of expensive or limited infrastructure supply, the poor are often forced to find alternatives to utility suppliers by illegally tapping into the "formal city." In Rio, *gatos* (literally "cats"), is the term for the unofficial and often illegal connections made to water and electricity infrastructures. As Fabricius argues, *gatos* are emblematic of the relationship between formal and informal urban areas, and "represent the point of transition from a controlled and regulated network to an 'invisible' and undocumented complex."[625] *Gatos* are often the only means through which communities can gain access to the city's infrastructures. Regularization of settlements can be extremely slow and itself does not necessarily entail infrastructure provision. As Fabricius points out, these complex and fragile infrastructure connections are highly visible, unlike those of the

Figure 9.2 Infrastructure billboard, Mumbai. Source: authors collection.

formal city (see Figure 9.3 from Mumbai). Infrastructure provision here is a form of bricolage, with connections emplaced often literally one house and lane at a time, positioning improvisation and a range of intermediary providers as key partners to interruption.

This process of improvisation is a necessary and continuous effort. As Coelho writes of water provision in Chennai: "Through collusions between frontline engineers and the public, and in the easy unofficial modes so typical of official practice, bypass connections were effected, valves were manipulated, furtive handpumps were installed, pipes were raised and lowered."[626] Gandy argues in relation to Lagos that the intense social polarization and spatial fragmentation in the city since the mid-1980s have led to a scenario in which many households, rich, and especially poor, improvise their own water supply, power generation, and security services, leading to the emergence of a "self-service city" with its own daily rhythm:

As night falls, the drone of traffic is gradually displaced by the roar of thousands of genera-

tors that enable the city to function after dark. Many roads in both rich and poor neighbourhoods become closed or subject to a plethora of ad hoc check-points and local security arrangements to protect people and property until the morning.[627]

Less than 5 percent of households in Lagos have access to piped water, and even these more fortunate people must contend with regular interruptions caused by power outages.

The level of connection households have to different infrastructures tends to reflect the relative cost and importance of a particular infrastructure to a household. Within informal settlements, water is usually the first form of connection sought given its paramount importance and given that it can usually be arranged relatively cheaply in the form of communal standpipes. Electricity tends to follow, while sewerage (the most expensive connection) and telephone connections tend to follow much later, if at all (see Table 9.1).[628] These different infrastructures are enrolled in the geographies of inequality in the city. Describing a vivid picture of splintering urbanism and "concrete divide" in Lagos, Gandy argues that "the prohibitive capital costs of improvements in water and sanitation infrastructure help to explain the growing 'concrete divide' between the rapid spread of new telecommunications technologies throughout the city in the context of widening inequalities in social and environmental conditions."[629] In the context of profound inequalities in water and sanitation infrastructures, Nigeria has the fastest growing mobile phone use in Sub-Saharan Africa.

During the 1990s, particularly in relatively wealthier cities in the Global South, there was a significant increase in infrastructure provision to the levels shown in Table 9.1, but this provision was highly unequal and often served to entrench existing urban inequalities. In urban China, for example, since the economic liberalization of the 1990s, selected buildings and gated enclaves have begun bypassing municipal water supply systems through the construction of small-scale secondary pipe networks for purified water supplies, giving rise to a dual water system.[630] Beyond the lack of water connections and the quality of the water, a key problem for the urban poor is regularity of access to clean water, of which there is enormous variability over time. In Sao Paulo, 82 percent of all households experience interruptions in the water supply every day.[631] In one survey of seven Indian cities, "about 18 per cent of the households who reported using municipal tap water supply stated that the tap water supply was available for 24 hours, while about 25 per cent and 27 per cent claimed that it was available for a few hours (less than 4 hours) twice a day, and once in a day, respectively."[632] In an effort to cope with a highly erratic supply, the urban poor often store water in drums and tanks. Shaban and Sharma echo the concerns of many campaigners in calling for a comprehensive

Figure 9.3 Fragile infrastructures. Godiwala, Khar, Mumbai, 2006. Source: author's collection.

national water policy to replace the current "adhocism."[633]

In response to this pattern of interruption, the World Bank has been arguing for some time that Indian cities have water shortages and interruptions precisely because the pricing of water is below costs.[634] The Bank has called for privatization and increased user charges in a great number of cities. In practice, more hybrid forms of state–commercial practices are emerging. For example, Gilbert's work on water supply in Bogota, Columbia, narrates how the water company remains in public hands but has increasingly followed more commercialized practice and has subcontracted various functions to the private sector. For Gilbert, the case of Bogota signals a note of caution to both neoliberals and anti-neoliberals alike because it demonstrates the abilities of water companies to

Table 9.1 Connections to Infrastructure: Regions and Informal Settlements (percentage).

Region	Water connection	Access to water	Sewerage	Electricity	Telephone
Sub-Saharan Africa	R: 48.4	R: 73.5	R: 30.9	R: 53.9	R: 15.5
	I: 19.1	I: 40	I: 7.4	I: 20.3	I: 2.9
North Africa and the Middle East	R: 79.1	R: 88	R: 65.9	R: 91.8	R: 42
	I: 35.7	I: 42.7	I: 21.5	I: 35.9	I: 30
Asia and the Pacific	R: 65.9	R: 94.8	R: 58	R: 94.4	R: 57.1
	I: 38.3	I: 89.1	I: 7.4	I: 75.7	I: 25.4
Latin America and the Caribbean	R: 83.7	R: 89.1	R: 63.6	R: 91.2	R: 51.7
	I: 57.9	I: 66.8	I: 30.3	I: 84.7	I: 32
Across global South	75.8	88.9	64	86.5	52.1
Across global South (informal settlements)	37.2	57.6	19.8	59.1	25.4

Note: R = region; I = informal settlements. *Water connection* refers to percentage of households with a piped water connection. *Access to water* means having potable water within 200 meters of the household (e.g., standpipes, wells, etc.) and includes water connections (since most countries presume that piped water is potable). The data for informal settlements may contain inaccuracies because sample sizes are small and measurements are uncertain.[30]
Source: UN Habitat (2003: 113–114).

combine elements of traditionally public and private practices:

> The experience of Bogota cannot tell us whether similar approaches can work anywhere else, but it does demonstrate that commercial practice combined with relative efficiency, political autonomy and, very importantly, a commitment to cross-subsidies can create a public company that actually delivers. Both rich and poor receive a generally good and reliable service, the rich contribute to the service for the poor and only the extremely poor struggle to pay their bills.[635]

However, as water connections are increasingly metered and charges increase across cities as has been recently witnessed in Delhi, it is likely that interruption in water connections will increasingly result from nonpayment by users due to high charges. This issue of nonpayment is, of course, politicized, and states can use nonpayment as a reason for cutting supplies to cover for failures in repair and maintenance, as well as for rerouting supplies. This is relevant, for example, to the political debate around water supply in urban South Africa, and the increased media reporting about water cut-offs. Reports

claimed that under the government's 1996 Growth Employment and Redistribution (GEAR) macroeconomic policy, poor peoples' recently acquired access to water had been lost when they were cut off for "nonpayment."[636] However, a household survey across urban South Africa investigated the extent and perceived cause of interruptions in water supply.[637] Muller writes:

> When asked "In the past year how often did you experience interruptions of the water service?" 2.6 percent of respondents reported interruptions "at least once a month" and a further 15 percent "several times a year." However, only 7 percent of interruptions were reported to be "for non-payment." Instead, 38.9 percent reported that they were interrupted for repairs and 39.3 percent that "it just stopped."[638]

While such data has its own irregularities, municipalities are not immune from shirking responsibility for technical failure by claiming nonpayment as the reason for cut-offs. While the nature of and responses to interruption vary considerably, the relations between infrastructure and inequality are central to the perception and experience of interruption across cities in the Global South. In the rest of the chapter, I explore this in more detail by considering infrastructure crisis.

INFRASTRUCTURE CRISIS AND URBAN INEQUALITY

The inequalities woven through urban infrastructures are rarely more evident or visible than in times of crisis or rupture. When water, energy, or transport networks suffer extreme pressures or collapse completely, the underlying urban power geometries can become clearly perceptible.[639] There are three aspects to this that I will raise in the remainder of the chapter which draw attention to the relations between infrastructure, inequality, and interruption.[640] First, in the case of largely unforeseen crises, it is often in the *responses* of local governments and infrastructure managers that we can analyze the reinforcement or transformation of power relations. Second, crises are mediated by and often exacerbate existing forms of inequality. Dominant groups have the capacities to insulate themselves from infrastructure collapse, and to reinforce their positions and interests through their discursive mobilization of crises and material adaptation to extreme conditions. Giglioli and Swyngedouw, for example, focus on the sociopolitical relations underpinning water supply in Sicily and the discursive and material construction of scarcity during the Sicilian water crisis of summer 2002.[641] They explore how hegemonic groups—constituted between organized crime and powerful politicians around the local–national nexus—maintain control over water infrastructure, and how these techno-natural networks become the basis for debates around the rearticulation of a (corrupt and clientelistic) hegemonic power base. Infrastructure crisis here becomes an explicit political instrument, legitimating political-economic "speculation on thirst."

Third, these forms of "crisis" are becoming increasingly common, raising the question of whether they actually constitute a "new normality" for some groups.[642] The overcapacity and underutilization of water and wastewater infrastructures in Eastern Germany resulting from "shrinking" processes of parallel population and economic decline, appears to Moss to be more of a "new normality" than a temporary problem which will resolve itself in time. Giglioli and Swyngedouw analyze an evolving political situation that is shaped within the normal, habitual context of drought and (perceived) water scarcity in Sicily, albeit one that "exploded" dramatically into view during an exceptional summer. The insertion of seemingly extreme events into the fabric of everyday life positions infrastructure in a wider regime of asset and liability, hazard and risk, and inertia and failure.

This is powerfully illustrated by Humphrey, who, writing about infrastructure in the eastern Siberian city of Ulan-Ude, shows how public perceptions of infrastructure shifted from a notion that it was a "taken-for-granted" foundation to urban life to a sense of infrastructure as threat, as malign in the hands of particular corporate and state interests.[643] In the Soviet context, the notion of infrastructure as foundation is distinct from the more Western perception of infrastructure as a stage upon which life takes place.[644] It is bound up with a Marxist understanding of *infrastruktura*—particularly prominent in the early years of Soviet rule—as, like ideology, a determining force both of the level of development and modernity, and the level of societal consciousness. As Humphrey argues, "a lot was at stake with the infrastructure so conceived," and indeed breakdown takes on a particular social resonance that differs from other urban contexts.[645] For example, in addition to the close wrapping of infrastructure with Leninist ideas of urban development and consciousness, the highly spread-out nature of Siberian cities, combined with harsh winters, means that it matters a great deal when infrastructures like energy break down.

Even if infrastructure provision in Siberian cities continues to follow the traditional Soviet state-led model with no legal obligation to pay either for access or consumption, for Humphrey, infrastructure for many people is increasingly a "source of anxiety and destabilization."[646] Debates and privatization often take place in contexts of anxiety and decay. There is strong evidence that across the global North–South divide we are witnessing a general corporatization of urban infrastructure, maintenance, and repair.[647] In poorer cities, a key driver of this process has been structural adjustment, which has reduced the role of state investment in infrastructure and pressured states to deregulate markets and privatize services like water and electricity.[648] The impetus of many cities to recreate themselves as tourist or convention cities is part of this process, leading often to changing land values and socioeconomic displacement of poorer and ethnic groups. In addition, infrastructure repair and recovery can expose cities in new ways to neoliberal capitalist pressures.

Writing about what they call "wounded cities," Schneider and Susser's edited collection shows how large-scale corporate developers target damaged landscapes, often taking advantage of tax breaks. As Harvey has argued, the capacity of cities to overcome a range of large and small tribulations is in part a result of the role of capitalist cities as hyperactive sites of "creative destruction": "They dance to the capitalist imperative to dismantle the old and give birth to the new as expanding capital accumulation accompanied by new technologies, new forms of organization and rapid influxes of populations (now drawn from all corners of the earth) impose new spatial forms and stresses upon the physical and social landscape."[649] One powerful illustration of the three relations between crisis, infrastructure, and inequality highlighted above comes from the monsoon floods that devastated thousands of lives in Mumbai in 2005 and which reflected the city's particular trajectory of "disaster capitalism," to which the chapter now turns.[650]

URBAN FLOODING IN MUMBAI

On July 26, 2005, 944 mm of rain fell in a five hour period in the city of Mumbai, creating floods that covered one-third of the city's surface and which reached almost 5 meters in depth in low-lying areas. This was the heaviest rainfall since records began in 1846, but there was no weather warning. Electricity supplies were cut, mobile phone networks faltered, public transport ground to a halt, and the city's suburban rail system—key to the economy and social fabric—was out of service for eighteen hours. The drainage system was overwhelmed, and in places floodwaters did not recede for days, leading to localized outbreaks of malaria, dengue, and leptospirosis. Over one thousand people were estimated to have been killed in the destruction, predominantly in informal settlements—drowned, electrocuted, or buried in landslides. The government imposed a mandatory holiday, ordering all workers to stay at home, and closed the stock exchange and city banks. Many people spent up to two days away from home, trapped on roofs or in schools with little or no food, water, or medicines. In one settlement in Govandi, north Mumbai, a toilet block, as the highest structure in the area, provided refuge for local people as floodwaters washed away or destroyed fragile housing and infrastructure.

In interviews I conducted in an informal settlement in Khar, Bandra in 2006, local people, having lost property and suffered housing and infrastructure damage, were fortunate in that they actually received promised government help—Rs. 5,000 for each family through central government relief, as well as distributions of rice, wheat, dahl, and cooking oil—but had to wait a month for it. One resident, Vasin-al-Hassan Qureshi, lamented: "We lost everything. Our furniture, clothes.

Nothing is remaining here." He had walked 45 km from Govandi to his house, having spent the first night standing on a flyover at Kurla. In a nearby settlement in Santa Cruz, a western suburb that received higher levels of flooding, one woman who suffered two meters of flooding waited three months for the Rs. 5,000 compensation, and complained that many families received nothing and that donations were often dependent on government connections rather than actual need.

This claim was consistently repeated. One man said people had given up trying to get compensation: "Why go fight with the government? We cannot go fight with the government for not doing things they should have done." In Shivaji Nagar, Govandi, a sewer line running through the center of the informal settlement blocked during the monsoon flooding and spilt onto the lanes. Apnalaya, a local NGO working in the area, gave kerosene tablets and medication for jaundice. Parts of the settlement were practically washed away by garbage in the flood. In the following days, people were left living on piles of newspaper or under plastic sheets supported by a bamboo stick, on a nearby city dumping ground. In terms or "rehabilitation," Apnalaya said there were no compensation payments, no grain, and no kerosene from the state, partly because the settlement is unauthorized and partly because of a high Muslim population in the area, which has historically suffered from state prejudice.

On the days of the floods, there were no back-up systems for power supplies, no wireless communication between key agencies such as police, fire, or hospitals, no effective public information systems, and widespread failure of call center operations because of poor Internet connectivity. As many commentators have shown, the state authorities were in disarray following the floods and largely abdicated responsibility. But, as Anjaria points out, although the police and municipal government were nonfunctional, there was no looting, theft, physical assaults, or

sexual violence: "Instead, the public showed a huge outpouring of spontaneous acts of kindness and generosity. Indeed, despite the wide scale destruction and infrastructural failure, something in the city worked."[651] Media reports were full of stories of "slum dwellers" rescuing those stranded in cars, offering chai and biscuits, and in many cases a room to sleep in. In contrast to discourses of social collapse and malaise, the city demonstrated an infrastructure of generosity and hospitality in the face of severe and multiple network infrastructure collapse.

The event led to a wide-ranging debate amongst various publics in the city about the relations between infrastructure, state responsibility, and the possibilities of a modern city free from infrastructure failure and interruption. The Concerned Citizens' Commission (CCC), a network of environmental activists and journalists, argued that a combination of uncontrolled development over wetland areas—including the destruction of mangroves to make way for new property development, or construction along the city's Mithi River—illegal dumping and "encroachment" on storm water drains, a lack of government planning for disasters, and the simple combination of unprecedented rainfall and high tide in a low-lying city, were key causes.[652] According to the CCC report, two large-scale development projects—the expansion of a runway for the international airport and massive land reclamation for the Bandra-Kurla complex (a new business district)—narrowed the river's width as well as radically altering its course. As a result, and as the findings of the CCC show, people living in these areas experienced the worst effects of the flooding.[653] As Anjaria writes:

> The worse incident occurred in Saki Naka, where a section of a hill collapsed, causing dozens of shanties to collapse, killing over a hundred people. In other areas in the suburbs, less sturdily built shanties were simply washed away along with their contents, which sometimes included small children.[654]

The media sought explanations for the extent of flooding, but only rarely mentioned the important Brihanmumbai Storm Water Drainage (Brimstowad) report, which was produced in response to flooding from the 1985 monsoon floods and which had outlined the need for major infrastructural improvements to the storm water drainage network. The improvements were stalled through a combination of a lack of state investment, delayed projects, and the state's inability to satisfy World Bank loan conditions. Much more focus in the media was given to disaster planning and the implementation of laws, for instance legal frameworks regulating where development projects can take place. For example, the Coastal Regulation Zone rules prevent construction on the city's coast. However, development can take place on the noncoastal side of coastal roads, and development plans often identify coastal roads that never exist in reality in order to allow construction.[655] One journalist at the time wrote provocatively of the "rape of infrastructure" by developers and the state, and pointed out that the floods encouraged debate on the relationship between builders, developers, and the state on the one hand, and sustainability on the other, even if it ultimately led to little change in those relations.

Blame for the flooding was assigned to a variety of causes: "slums" blocking drains, and developers blocking drains with construction debris; lack of state investment in drainage; rampant, uncontrolled development resulting in loss of mangroves, river space, and the natural drainage of green spaces; a freak episode of rainfall; rising sea level and climate change; poor state planning for disaster management and ability or commitment to respond to crisis; and flouting of construction regulations. The attachment of blame oscillated in public debates between the state, the poor, and "nature" broadly conceived, and clearly there were political advantages to who assigned blame to what. The state blamed unprecedented rainfall and pleaded that it could

not possibly have prepared for or responded to the event. In interviews I conducted with engineers at the city's Storm Water Drainage department, engineers consistently defended themselves by claiming that the monsoon had been a "once in 1000 years" event. One engineer claimed: "When high tide is there, no one can prevent flooding," and when asked who should be accountable, he laughed: "See, flooding is inevitable." At times the state became more desperate, and even blamed plastic bags for the floods. Maharashtra Chief Minister Vilasrao Deshmukh defiantly stated following the event: "Discarded plastic bags choked the drains and the sewerage and were mainly responsible for the unprecedented waterlogging."[656] The state's public works department began construction of a new sea wall stretching from Colaba to Gorai. Some elements within the state, and more conservative city environmental groups, such as Citispace, blamed the poor for blocking up storm drains with "illegal encroachment," and called for renewed efforts at demolition. One engineer claimed that the "success of our project depends on these [slum] hutments" being removed, and complained that a liberal culture in the city pervaded which was "for slums," meaning informal settlements could not be held responsible for infrastructure blockages.

While the levels of rainfall were indeed unprecedented, the state was exercising simple political expediency in blaming "nature" for the event, and familiar bourgeois prejudice in blaming the poor. The state is responsible for the catastrophic loss of life and damage to housing, property, and infrastructure, which was an indirect result of decades of lack of investment in drainage, flouting of building regulations, and uncontrolled construction (not by the poor, but by developers in collaboration with state officials), and a lack of disaster planning and coordination. However, given that "nature" is an unwieldy target, it was the poor that shouldered much of the blame for the floods as the alleged cause of the blocking of the storm drainage infrastructure. The Chitale commission, a state investigation into the floods focused on the part of Mumbai around the airport and the Mithi River, published its findings in 2006, and within days the state announced a new three-year deadline for slum clearance around the international airport. A similar rationale has been given for demolition in other parts of the city.

The relations between the urban imagination, modernity, and infrastructure were altered as a result of the event. Politico-corporate Mumbai's fantasies of transforming the city into the "new Shanghai" with "world-class" infrastructure fell apart amid the scenes of an overwhelmed city.[657] In this sense, the Mumbai floods represent the converse of the spectacle of Delhi's new metro. One of India's leading business dailies, the *Economic Times*, wrote: "Any aspirations, illusions or fine daydreams for standing up to some of the biggest and best global cities have now been crashed into smithereens."[658] Moreover, the event caused people to question whether this form of devastating

interruption could become a new norm for Mumbai. The following year, in anticipation of yet more floods in the face of slow progress with storm water drain improvements and continued urbanization of natural drainage and coastal spaces, the city's popular tabloids printed frequent stories and images that echoed the engineers' resignation to flooding (see Figure 9.4).[659]

As with many unforeseen infrastructure crises,[660] the 2005 monsoon floods starkly revealed that it is often in the responses of local governments that we can analyze the reinforcement or transformation of power relations. The abdication of state responsibility and slow delivery of relief and compensation reveals an urban state unable to respond to the most basic needs of the majority of its inhabitants. However, the public response to the event led to a wide-ranging discussion of the causes of the floods, and linked Mumbai's form of urban development to social inequality and environmental sustainability. In doing so, a renewed emphasis was placed on the politicization of urban infrastructure. The monsoon crisis exacerbated existing

Figure 9.4 Monsoon floods, Mumbai, February, 2006. Source: Mumbai Mirror, February 6th, 2006.

forms of inequality in terms of escalating demolition of informal settlements, but in the same instance led to a focus on the limits and possibilities of infrastructure and development. As monsoon flooding becomes an ever more familiar sight in particular parts of Mumbai, however, this particular form of "crisis" and its myriad effects are becoming a new normality for some groups, especially the urban poor.

CONCLUSION

This chapter has argued that infrastructure reflects and reproduces urban inequality. Interruption and crisis can lay bare this inequality. In particular, the responses to infrastructure crisis, and the fact that certain forms of crisis have or are becoming a normality for some groups of (especially) the urban poor, can starkly reveal urban inequality. The urban poor cope with interruption primarily through improvisation that draws on a variety of intermediaries, while states increasingly opt for investment in high-end infrastructures that may improve the attractiveness of the urban fabric to investment capital. In the face of poor quality and interrupted services, the increasing tendency of states toward privatizing basic services such as water and electricity infrastructures, whether by choice or through pressure from agencies like the World Bank, raises the prospect of increased costs and continued illegal tapping of infrastructures for the poorest.

But despite the undoubted prevalence of urban infrastructure interruption, the characterization of cities in the Global South as synonymous with breakdown or failure ignores the complexity of urban life and recasts a developed/functional versus developing/dysfunctional dichotomy. This dichotomy does not help to move beyond the now staid debates on "slum" as either symbol of urban failure or malaise, or as skillful space of entrepreneurialism. There is a need for further work that extends the conceptions of the relations between informal settlements, infrastructure, and agency on an everyday basis. This effort could also address the relative lack of research on how different groups perceive experience and respond to everyday forms of infrastructure interruption.

NOTES

1. G. Deleuze and F. Guattari, "City/State," *Zone* 1/2(1997): 196.
2. United Nations Habitat Program, *State of the World's Cities* (Nairobi, Kenya: United Nations Habitat Program, 2007), 4.
3. See O. Coutard, ed., *The Governance of Large Technical Systems* (London: Routledge, 1999).
4. See, for example, P. Steinberg and R. Shields, eds., *What Is a City? Rethinking the Urban after Hurricane Katrina* (Athens and London: University of Georgia Press, 2008).
5. R. Herman and J. Ausubel, *Cities and Their Vital Systems—Infrastructure Past, Present and Future* (Washington, DC: National Academy Press, 1988).
6. See D. Sudjic, *The Edifice Complex: How the Rich and Powerful Shape the World* (London: Penguin, 2006): 1.
7. S. Graham and S. Marvin, "Telematics and the Convergence of Urban Infrastructure: Implications for Contemporary Cities," *Town Planning Review* 65, no. 3 (1994): 227–42.
8. G Strand, "Keyword: Evil," *Harper's Magazine*, March 2008, 64–65.
9. S. Graham, "Beyond the 'Dazzling Light': From Dreams of Transcendence to the 'Remediation' of Urban Life," *New Media and Society* 6, no. 1 (2004): 16–26.
10. M. Benedikt, *Cyberspace: First Steps* (Cambridge, MA: MIT Press, 1992): 12.
11. B.Carroll, "Seeing Cyberspace: The Electrical Infrastructure as Architecture," in *Sarai Reader: Cities and Everyday Life* (New Delhi: Sarai, 2002), http:// www.sarai.net/publications/readers/02-the-cities-of-everyday-life/04cyber_ electric.pdf (accessed January, 2009).
12. Ibid., 250.
13. G. Strand, "Keyword: Evil," 64–65.
14. D. Perry, "Introduction," in D. Perry (Ed.) *Building the Public City: The Politics, Governance and Finance of Public Infrastructure* (London: Sage, 1995), 1–20.
15. Steve Hinchliffe defines black boxes "as settled items whose users and colleagues (human and non-human) act in ways which are unchallenging to the technology." S. Hinchliffe, "Technology, Power and Space—The means and ends of geographies of technology," *Environment and Planning D: Society and Space* 14 (1996): 665.
16. B. Mau, *Massive Change* (London Phaidon, 2003), 3–4.
17. Ibid., 3–6, 8.
18. G Bowker and S. L. Star, *Sorting Things Out: Classification and Its Consequences* (Cambridge, MA: MIT Press, 2000).
19. S. L. Star, "The Ethnography of Infrastructure," *American Behavioral Scientist* 43, no. 3 (1999): 379.
20. G. Bowker and S. L. Star (*Sorting Things Out: Classification and Its Consequences* (Cambridge, MA: MIT Press, 2000), 33.
21. Bowker and Star use seven other characteristics to define the term *infrastructure*. To them, infrastructure is embedded (i.e., "sunk into other structures"); transparent ("it does not need to be

reinvented each time or assembled for each task"); offers temporal or spatial reach or scope; is learned by its users; is linked to conventions of practice (e.g., routines of electricity use); embodies standards; is built on an installed base of sunk capital; and is fixed in modular increments rather than being built all at once or globally (Bowker and L. Star, 335).

22. See A. Ong and S. Collier, eds., *Global Assemblages: Technology, Politics, and Ethics as Anthropological Problems* (Oxford, UK: Blackwell, 2005).

23. M. Kaika and E. Swyngedouw, "Fetishising the Modern City: The Phantasmagoria of Urban Technological Networks," *International Journal of Urban and Regional Research* 24, no. 1 (2000): 122.

24. Ibid., 122.

25. See, for example, D. Pike, *Metropolis on the Styx: The Underworlds of Modern Urban Culture, 1800–2001* (New York: Columbia University Press, 2007).

26. See, for example, J. Solis, *New York Underground* (New York: Routledge, 2007).

27. J. Summerton, ed., *Changing Large Technical Systems* (Boulder, CO: Westview Press, 1994), 1.

28. R. Mayntz, "Technological Progress, Societal Change and the Development of Large Technical Systems." Mimeo, 1995, 5.

29. Summerton, *Changing Large Technical Systems*, 1.

30. Two recent collections of papers provide excellent explorations of the multiple relations between networks and infrastructures, degrees of black boxing by users, and the politics of disruptions. See the eleven papers within Oliver Coutard's special issue of *Geoforum*, "Placing Splintering Urbanism," 39, no. 6; and the papers edited by Colin McFarlane and Jonathan Rutherford in the special issue of *International Journal of Urban and Regional Research* 32, no. 3 (2008).

31. See, for example, S. Speak and S. Graham, "Service Not Included: Marginalised Neighbourhoods, Private Service Disinvestment, and Compound Social Exclusion," *Environment and Planning A* (2000): 1985–2001.

32. C. McFarlane and J. Rutherford, "Political Infrastructures: Governing and Experiencing the Fabric of the City," *International Journal of Urban and Regional Research* 32, no. 2 (June 2008): 363–74.

33. See S. Graham and S. Marvin, *Splintering Urbanism* (London: Routledge, 2001) for an example of a synoptic book which has this tendency.

34. The case of Mumbai is instructive here. See S. Chatterjee, "Mumbai's 'Shanghai Dream' Hangs in Balance," at Bombay's Demolition Drive, blog February 27, 2005, http://dupb.blogspot.com/2005_02_01_archive.html.

35. McFarlane and Rutherford, "Political Infrastructures," 363–74.

36. J. Summerton, "Social Shaping In Large Technical Systems," *Flux*, 17 (July–September1994): 56.

37. W. Rammert "New Rules of Sociological Method: Rethinking Technology Studies," *British Journal of Sociology*, 48(1997): 171–91.

38. See S. Graham and N. Thrift, "Out of Order: Understanding Maintenance and Repair," *Theory, Culture and Society*, 24, no. 3(2007): 1–25.

39. Department for Professional Employees (DPE), *Current Statistics on White Collar Employees* (Washington, DC: DPE, 2003).

40. Graham and Thrift, "Out of Order," 2.

41. See, for example, McFarlane and Rutherford, "Political Infrastructures," 363–74.

42. See, for example, J. Urry, *Mobilities* (Cambridge, UK: Polity, 2007); T. Cresswell, *On the Move* (London: Routledge, 2006).

43. See B. Latour, *We Have Never Been Modern* (London: Harvester and Wheatsheaf, 1993); John Law and John Hassard, eds., *Actor Network Theory and After* (London: Wiley, 1999).

44. See A. Ong and S. Collier, eds., *Global Assemblages: Technology, Politics, and Ethics as Anthropological Problems* (Oxford, UK: Blackwell, 2005). Following Gilles Deleuze, Jane Bennett emphasizes that: "an assemblage is, first, an ad hoc grouping, a collectivity whose origins are historical and circumstantial, though its contingent status says nothing about its efficacy, which can be quite

strong. An assemblage is, second, a living, throbbing grouping whose coherence coexists with energies and countercultures that exceed and confound it. An assemblage is, third, a web with an uneven topography: some of the points at which the trajectories of actants cross each other are more heavily trafficked than others, and thus power is not equally distributed across the assemblage. An assemblage is, fourth, not governed by a central power: no one member has sufficient competence to fully determine the consequences of the activities of the assemblage. An assemblage, finally, is made up of many types of actants: humans and nonhumans; animals, vegetables, and minerals; nature, culture, and technology." J. Bennett, "The Agency of Assemblages and the North American Blackout," *Public Culture* 17, no. 3 (2005): 445.

45. Ibid., 446.
46. Ibid.
47. N. Heynen, M. Kaika, and Erik Sywyngedouw, eds., *In the Nature of Cities: Urban Political Ecology and the Politics of Urban Metabolism* (London: Routledge, 2006).
48. M. Kaika and E. Swyngedouw, "Fetishising the Modern City, 122.
49. See M. Kaika, *City of Flows* (London: Routledge, 2004).
50. See Heynen et al., *In the Nature of Cities*.
51. M. Gandy, "Cyborg Urbanization: Complexity and Monstrosity in the Contemporary City," *International Journal of Urban and Regional Research* 29, no. 1 (March 2005): 26–49. See also M. Gandy, *Concrete and Clay: Reworking Nature in New York City* (Cambridge, MA: MIT Press, 2002).
52. See C. Hables Gray, *The Cyborg Handbook* (New York: Routledge, 1995); D. Haraway, "A Manifesto For Cyborgs: Science, Technology, and Socialist-Feminism in the Late Twentieth Century," in *Simians, Cyborgs and Women: The Reinvention of Nature* ed. D. Haraway, (New York: Routledge, 1991), 149–81.
53. See Gandy, "Cyborg Urbanization," 26–49; and Antoine Picon, *La Ville Territoire de Cyborgs* (Paris: L'Imprimeur, 1998).
54. Gandy, "Cyborg Urbanization," 28. See also M. Kaika, "Interrogating the Geographies of the Familiar: Domesticating Nature and Constructing the Autonomy of the Modern Home," *International Journal of Urban and Regional Research*, 28, no. 2 (2004): 265–86.
55. G. Bateson, *Steps to an Ecology of Mind* (New York: Ballantine, 1978), 249.
56. Mcfarlane and Rutherford, "Political Infrastructures," 363.
57. S. L. Star, "The Ethnography of Infrastructure," *American Behavioral Scientist* 43, no. 3 (1999): 377.
58. See D. Harvey, *The Urbanization of Capital* (Oxford. UK: Blackwell, 1985); and E. Swyngedouw, "Communication, Mobility and the Struggle for Power Over Space," in *Transport and Communications in the New Europe*, ed. G. Giannopoulos and A. Gillespie, 305–25 (London : Belhaven, 1993).
59. D. Harvey, *The Urbanization of Capital* (Oxford, UK: Blackwell, 1985).
60. W. Bijker, "Do Not Despair: There Is Life After Constructivism," *Science, Technology and Human Values* 18, no. 1 (1993): 113–38.
61. Gandy, "Cyborg Urbanization," 26–49.
62. Star, "Ethnography of Infrastructure," 380.
63. See S. Graham, "Software-Sorted Geographies," *Progress in Human Geography* 29, no. 5 (2005): 562–80.
64. See M. Salter, "The Global Visa Regime and the Political Technologies of the International Self: Borders, Bodies, Biopolitics," *Alternatives* 31 (2006): 167–89.
65. See J. Garvey, *The Ethics of Climate Change: Right and Wrong in a Warming World* (London: Continuum, 2008).
66. See S. Graham, *Cities Under Siege: The New Military Urbanism* (London: Verso, 2009), chap. 10.

67. See S. Branford, "Food Crisis Leading to an Unsustainable Land Grab," *The Guardian*, November 22, 2008. http://www.guardian.co.uk/environment/2008/nov/22/food-biofuels.

68. See Graham, *Cities Under Siege,* chap. 10.

69. See S. Graham and S. Marvin, *Splintering Urbanism: Networked Infrastructure, Technological Mobilities and the Urban Condition* (London: Routledge, 2001).

70. S. Graham, "Constructing Premium Networked Spaces: Reflections on Infrastructure Networks and Contemporary Urban Development," *International Journal of Urban and Regional Research,* 24, no. 1 (2000): 183–200.

71. See M. Torrance, "Forging Glocal Governance? Urban Infrastructures as Networked Financial Products," *International Journal of Urban and Regional Research* 32, no. 1 (2008): 1–21.

72. See N. Klein, *The Shock Doctrine: The Rise of Disaster Capitalism,* (New York: Penguin, 2007).

73. G. Rochlin, "Networks and the Subversion of Choice: An Institutionalist Manifesto," *Journal of Urban Technology* 8, no. 3 (2001): 65–96.

74. Ibid.

75. J. Bennett, "The Agency of Assemblages and the North American Blackout," *Public Culture* 17, no. 3 (2005): 463.

76. Ibid., 463–64.

77. See L. Vale and T. Campanella, eds, *The Resilient City: How Modern Cities Recover From Disaster* (Oxford, UK: Oxford University Press, 2005); J. Coaffee, D. Murakami Wood, and Peter Rogers, *The Everyday Resilience of the City: How Cities Respond to Terrorism and Disaster* (London: Palgrave Macmillan, 2008).

78. For a review, see A. Lakoff, "Preparing for the Next Emergency," *Public Culture* 19, no. 2 (2007): 247–71.

79. J. Reid, "Open Source Destruction, Weak Discipline, War Infrastructure," in *Under Fire.* ed., Jordan Crandall. http://underfire.eyebeam.org/?q=node/462 (accessed February 1, 2009).

80. E. Klinenberg, *Heat Wave: A Social Autopsy of Disaster in Chicago.* (Chicago: University of Chicago Press, 2006).

81. K. Easterling, *Enduring Innocence* (Cambridge, MA: MIT Press, 2005).

82. Ibid.

83. See G. Squires and C. Hartman, eds., *There Is No Such Thing as a Natural Disaster: Race, Class and Hurricane Katrina* (New York: Routledge, 2006).

84. Steinberg and Shields , *What Is a City?*

85. See J. Robb, "Cascading System Failure," http://globalguerrillas.typepad.com/ May 24, 2004 (accessed January 21, 2009).

86. S. Graham, "'Homeland' Insecurities? Katrina and the Politics of Security in Metropolitan America," *Space and Culture* 9, no. 1 (2006): 63–67.

87. See L. Amoore and M. de Goede, eds., *Risk and the War on Terror* (London: Routledge, 2007)

88. A. Feldman, "Securocratic Wars of Public Safety," *Interventions: International Journal of Postcolonial Studies* 6, no. 3 (2004): 333.

89. Ibid.

90. Ibid.

91. S. Barakat, "City War Zones," *Urban Age* (Spring 1999): 12.

92. Mau, *Massive Change,* 3–4.

93. Heideggerian philosophy and phenomenology is particularly helpful in explaining how the apparent failures of infrastructure systems work to render them visible. Such events "disclose a world," writes Peter-Paul Verbeek. "When somebody uses a tool or piece of equipment, a referential structure comes about in which the object produced, the material out of which it is made, the future user, and the environment in which it has a place are related to each other. But that this is so, according to Heidegger, generally appears only when a handy or ready to hand tool or piece of equipment breaks down. When this happens, the tool suddenly demands attention for itself. The reliable dealings we are used to having with the tool are ruptured, and instead of withdraw-

ing from our attention the tool suddenly forces itself upon us. Someone sits at a word processor focused on the text at hand and all of a sudden the computer freezes. The trustworthy world that developed around the computer—the open book, the keyboard, the screen, the cup of coffee; in short, the entire mutually referring network that Heidegger calls a world—is abruptly destroyed." (Peter-Paul Verbeek, *What Things Do: Philosophical Reflections on Technology, Agency and Design* (University Park, PA: Pennsylvania State University Press, 2004), 79–80, cited in Stephen Graham and Nigel Thrift, "Out of Order: Understanding Maintenance and Repair," *Theory, Culture and Society* 24, no. 3 (2007): 1–25.

94. See S. L. Star, "Ethnography of Infrastructure," 377–91.

95. See E. Goffman, *The Presentation of the Self in Everyday Life* (Garden City, NY: Doubleday, 1959); and C. Henke, "The Mechanics of Workplace Order: Toward a Sociology of Repair," *Berkeley Journal of Sociology* 43 (2000): 55–81. Rather than focusing on material infrastructure networks, Goffman, working within an ethnomethodological tradition, discussed the way "repair" operated as a kind of interactional maintenance of order in everyday encounters and conversations.

96. See R. Zimmerman "Social Implications of Infrastructure Network Interactions," *Journal of Urban Technology* 893 (2001): 97–119.

97. R. Little "Controlling Cascading Failure: Understanding the Vulnerabilities of Interconnected Infrastructure," *Journal of Urban Technology* 9, no. 1 (2002): 109–23.

98. C. Perrow, *Normal Accidents: Living With High-Risk Technologies* (Princeton, NJ: Princeton University Press, 1999).

99. Little, "Controlling Cascading Failure," 113.

100. J. Leslie, "Powerless," *Wired* (April 1999), 119–83. http://www.wired.com/wired/archive/7.04/blackout_pr.html.

101. Little, "Controlling Cascading Failure," 112.

102. See, for example, L. Vale and T. Campanella, eds., *The Resilient City: How Modern Cities Recover From Disaster* (Oxford, UK: Oxford University Press, 2005) and Jane Schneider and Ida Susser, eds., *Wounded Cities: Destruction and Reconstruction in a Globalized World* (London: Berg, 2005), 25–46.

103. See, for example, C. Sponster, "Beyond the Ruins: The Geopolitics of Urban Decay and Cybernetic Play," *Science Fiction Studies* 20, no. 2 (1992): 2251–265.

104. See, for example, M. Page, *The City's End* (New Haven, CT: Yale University Press, 2008); and Stephen Graham, ed., *Cities, War and Terrorism* (Oxford, UK: Blackwell. 2004).

105. See J. Diamond, *Collapse: How Societies Choose to Fail or Survive* (London: Penguin, 2006).

106. See, for example, R. Woodbridge, *The Next World War: Tribes, Cities, Nations, and Ecological Decline* (Toronto: University of Toronto Press, 2004).

107. W. Mitchell and A. Townsend, "Cyborg Agonistes: Disaster and Reconstruction in the Digital Era." in *The Resilient City: How Modern Cities Recover from Disaster* eds., L. Vale and T. Campanella, 313–24 (Oxford, UK: Oxford University Press, 2005).

108. See E. Ullman "The Myth of Order," *Wired*, 7, no.4 (1999). http://www.wired.com/wired/archive/7.04/y2k.html.

109. Star, "Ethnography of Infrastructure," 385.

110. See S. Graham and M. Crang, "Sentient Cities: Ambient Intelligence and the Politics of Urban Space," *Information, Communications and Society* 10, no. 6 (2008): 789–817.

111. M. Pawley, *Terminal Architecture* (London: Reaktion Books, 1997), 162.

112. W. Joy, "Why the Future Doesn't Need Us," *Wired* (April 2000): 238.

113. See, for example, R. Pineiro, *Cyberterror* (New York: Forge Press, 2003). For an excellent discussion of the way fears of "cyberterror" attacks are manipulated and exaggerated, see F. Debrix, *Tabloid Terror: War, Culture and Geopolitics* (New York: Routledge, 2008), chap. 1.

114. Graham and Marvin, "Telematics and the Convergence of Urban Infrastructure," 227–42.

115. Ullman, "Myth of Order," 126.

116. Ibid.,126.
117. Workforce Boards for Metropolitan Chicago (WBMC), *State of the Workforce* Report for the Chicago Metropolitan Region (Chicago: WBMC, 2003).
118. G. Rochlin, *Trapped in the Net: The Unanticipated Consequences of Computerization* (Princeton, NJ: Princeton University Press, 1997).
119. See K. Varnelis, "The Centripetal City: Telecommunications, the Internet, and the Shaping of the Modern Urban Environment," *Cabinet Magazine* 17 (Spring 2004), http://varnelis.net/articles/centripetal_city.
120. Ibid.
121. Ullman (1999), "Myth of Order," 126.
122. J. Konvitz, "Why Cities Don't Die," *American Heritage of Invention and Technology* (Winter 1990): 62.
123. W. Mitchell and A. Townsend, "Cyborg Agonistes," 313–24.
124. Schneider and Susser, *Wounded Cities,* 25–46.
125. See S. Redhead "The Art of the Accident: Paul Virilio and Accelerated Modernity," *Fast Capitalism* 2, no. 1 (2006). http://www.uta.edu/huma/agger/fastcapitalism/2_1/redhead.htm; and Paul Virilio, *Unknown Quantity* (London: Thames and Hudson, 2003).
126. Leslie, "Powerless," 119–83.
127. Ibid.
128. "Neighborhood #3 (Power Out)," from the album *Funeral.*
129. M. Sorkin, "Urban Warfare: A Tour of the Battlefield," in *Cities, War and Terrorism: Towards an Urban Geopolitics* ed. Stephen Graham, 259 (Oxford, UK: Blackwell, 2004).
130. Ibid.
131. J. Anjaria, "Urban Calamities: The View From Mumbai," *Space and Culture* 9, part 1 (January 2006): 80–92.
132. See, for example, B. Page, "Paying for Water and the Geography of Commodities," *Transactions of the Institute of British Geographers* 30, no. 3 (September 2005), 293–306.
133. See M. Davis, Buda's *Wagon: A Brief History of the Car Bomb* (London: Verso, 2007).
134. See J. Robb's influential global guerillas blog at http://globalguerrillas.typepad.com/ as well as his book *Brave New War: The Next Stage of Terrorism and the End of Globalization* (New York: Wiley, 2007).
135. See Graham, *Cities Under Siege.*
136. See V. Rao (2007), "How to Read a Bomb: Scenes from Bombay's Black Friday," *Public Culture* 19, no. 3 (2007): 567–92.
137. Ibid., 572.
138. S. Kostof, *The City Assembled* (London: Thames and Hudson, 1992), 305.
139. Little, "Controlling Cascading Failure."
140. Ibid.
141. See, for example, Sandia National Laboratory, http://www.sandia.gov/nisac/diisa.html.
142. Supervisory Control and Data Acquisition. SCADA refers to a system that collects data from various sensors at a factory, plant, or in other remote locations and then sends this data to a central computer which then manages and controls the data.
143. Perrow 1999a.
144. These occur where the systems involved are sufficiently complex to allow unexpected interactions of failures to occur such that safety systems are defeated, and sufficiently tightly coupled to allow a cascade of increasingly serious failures ending in disaster.
145. Perrow 1999b.
146. Bak and Paczuski 1995.
147. Talukdar et al. 2003.
148. Little 2004.
149. Amin 2003.

150. U.S.-Canada Power System Outage Task Force 2004.
151. Behr 2003.
152. Wong 1999.
153. Marburger 2002.
154. Wallace et al. 2003.
155. Zimmerman 2003.
156. O'Rourke, Lembo, and Nozick 2003.
157. Perrow 1999a
158. Sagan 1999.
159. La Porte and Consolini 1991.
160. Weick and Sutcliffe 2001.
161. La Porte 1994.
162. Perrow 1994.
163. La Porte and Rochlin 1994.
164. Weick and Sutcliffe 2001.
165. Petroski 1992, 1994.
166. Mileti 1999.
167. The ASRS is a voluntary program administered by NASA wherein air safety-related incidents and near accidents can be reported without fear of self-incrimination. The program is credited with facilitating beneficial change throughout the airline industry (Perrow 1999a).
168. Phimister et al. 2000.
169. IOM 2000.
170. Kletz 2001.
171. Ibid.
172. Taleb 2007.
173. ILIT 2006; IPET 2007.
174. Barry 1997.
175. Clarke 1999.
176. NRC 1996.
177. Kaplan and Garrick 1981.
178. Taleb 2007.
179. ILIT 2006; IPET 2007.
180. Haimes 1991.
181. New Orleans made use of specially designed dewatering pumps that were powered by 25 Hz alternating current which had to be produced close by the point of use. This led the generating capacity to be located within the area "protected" by the levees.
182. Anderson and Suess 2006.
183. *Insurance Journal* 2006.
184. Unless otherwise specified, newspaper articles referenced within this chapter were accessed between January 17 and February 9, 2006 on the NewsBank NewsFile Collection database, http://infoweb.newsbank.com.
185. C. E. Colten, *An Unnatural Metropolis: Wresting New Orleans From Nature* (Baton Rouge: Louisiana State University Press, 2005), 141.
186. Portions of this chapter draw on B. Sims, "'The Day After the Hurricane': Infrastructure, Order, and the New Orleans Police Department's Response to Hurricane Katrina," *Social Studies of Science* 37, no. 1 (February 2007): 111–18; and B. Sims, "Things Fall Apart: Disaster, Infrastructure, and Risk," *Social Studies of Science* 37, no. 1 (February 2007): 93–95. For an overview of science, technology, and society issues raised by the hurricane, see the other articles in this comments section on Hurricane Katrina in *Social Studies of Science* (February 2007).
187. J. Warrick, "Crisis Communications Remain Flawed," *The Washington Post*, December 10, 2005.

188. J. McQuaid, "Alarm Sounded Too Late as N.O. Swamped—Slow Response Left City in Lurch," *The Times-Picayune*, September 8, 2005.

189. S. Glasser and M. Grunwald, "The Steady Buildup to a City's Chaos," *The Washington Post*, September 11, 2005.

190. Warrick, "Crisis Communications."

191. D. Baum, "Deluged: When Katrina Hit, Where Were the Police?" *The New Yorker*, January 9, 2006.

192. Ibid.

193. Baum, "Deluged"; Joseph B. Treaster, "Police Quitting, Overwhelmed by Chaos," *New York Times*, September 4, 2005.

194. Baum, "Deluged"; G. Filosa, "N.O. Police Chief Defends Force," *Times-Picayune*, September 5, 2005.

195. Baum, "Deluged."

196. Baum, "Deluged"; M. Foster, "New Orleans Police Remain on Edge," *Houston Chronicle*, October 11, 2005.

197. Baum, "Deluged."

198. M. Douglas, *Purity and Danger: An Analysis of Concepts of Pollution and Taboo* (London: Routledge & Kegan Paul, 1966).

199. For example, see W. Bijker, T. Hughes, and T. Pinch, eds., *The Social Construction of Technological Systems: New Directions in the Sociology and History of Technology* (Cambridge, MA: MIT Press, 1987); D. MacKenzie, *Inventing Accuracy: A Historical Sociology of Nuclear Missile Guidance* (Cambridge, MA: MIT Press, 1990); W. Bijker, *Of Bicycles, Bakelites, and Bulbs: Toward a Theory of Sociotechnical Change* (Cambridge, MA: MIT Press, 1995); B. Latour, *Aramis, Or the Love of Technology* (Cambridge, MA: Harvard University Press, 1996).

200. Examinations of infrastructure from a science and technology studies perspective include S. Leigh Star and K. Ruhleder, "Steps Toward an Ecology of Infrastructure: Design and Access for Large Information Spaces," *Information Systems Research* 7, no. 1 (1996): 111–34; P. Edwards, "Infrastructure and Modernity: Force, Time, and Social Organization in the History of Sociotechnical Systems," in *Modernity and Technology*, ed. T. Misa, P. Brey, and A. Feenberg (Cambridge, MA: MIT Press, 2003), 185–225; and A. Hommels, *Unbuilding Cities: Obduracy in Urban Socio-Technical Change* (Cambridge, MA: MIT Press, 2005). On the history of infrastructure protection, see S. Collier and A. Lakoff, "The Vulnerability of Vital Systems: How 'Critical Infrastructure' Became a Security Problem," in *The Politics of Securing the Homeland: Critical Infrastructure, Risk and Securitisation*, ed. M. Dunn and K. Kristensen (London: Routledge, 2008).

201. B. Latour, "Where are the Missing Masses? The Sociology of a Few Mundane Artifacts," in *Shaping Technology/Building Society*, ed. W. Bijker and J. Law (Cambridge, MA: MIT Press, 1992), 225–58.

202. See Edwards, "Infrastructure and Modernity."

203. K. Weick, "The Collapse of Sensemaking in Organizations: The Mann Gulch Disaster," *Administrative Science Quarterly* 38 (1993): 633.

204. Ibid., 635.

205. T. Wachtendorf, "Improvising 9/11: Organizational Improvisation Following the World Trade Center Disaster" (PhD diss., University of Delaware, 2004); T. Wachtendorf and J. Kendra, "Improvising Disaster in the City of Jazz: Organizational Response to Hurricane Katrina," 2005, http://understandingkatrina.ssrc.org/Wachtendorf_Kendra. In his classic sociological study of disaster, *Everything in Its Path: Destruction of Community in the Buffalo Creek Flood* (New York: Simon and Schuster, 1976), Kai Erikson described the aftermath of a flash flood which physically obliterated several coal mining towns in West Virginia, while leaving many survivors. One of Erikson's more disturbing findings was that no sense of collective resilience or sense making emerged following that disaster, which also involved large-scale destruction of infrastructure.

206. C. Hustmyre, "NOPD Versus Hurricane Katrina," *Tactical Response Magazine*, July 2006; J. Arey

and A. Wilder, "NOPD SWAT Versus Katrina," *Law and Order Magazine*, January 2006; Baum, "Deluged"; M. Perlstein and T. Lee, "The Good and the Bad," *The Times-Picayune*, December 18, 2005.

207. In *Everything in Its Path*, Kai Erikson similarly found that disorientation in space and time and diminished commitment to the collective were key symptoms of a breakdown in social order.

208. B. Allen, "Environmental Justice and Expert Knowledge in the Wake of a Disaster," *Social Studies of Science* 37, no. 1 (February 2007): 103–10.

209. Douglas, *Purity and Danger*, 36.

210. Baum, "Deluged."

211. J. Treaster and J. DeSantis, "With Some Now at Breaking Point, City's Officers Tell of Pain and Pressure," *The New York Times*, September 6, 2005; Filosa, "N.O. Police Chief."

212. Baum, "Deluged."

213. A. Cha, "New Orleans Police Keep Public Trust, Private Pain," *The Washington Post*, September 12, 2005, at http://www.washingtonpost.com/wp-dyn/content/article/2005/09/11/AR2005091101460.html. (accessed, May 2009).

214. Treaster, "Police Quitting."

215. Foster, "New Orleans Police."

216. Baum, "Deluged"; M. Perlstein and T. Lee, "The Good and the Bad"; Hustmyre, "NOPD versus Hurricane Katrina."

217. J. Arey and A. Wilder, "New Orleans Police Respond to Katrina," *Law and Order Magazine*, October 2005.

218. Arey and Wilder, "NOPD SWAT Versus Katrina."

219. Baum, "Deluged."

220. Ibid.

221. Ibid.

222. Treaster and DeSantis, "With Some Now at Breaking Point."

223. Baum, "Deluged."

224. Cha, "New Orleans Police."

225. The Associated Press, "Katrina's Devastation Reaches into the Psyches of Survivors," *Pittsburgh Tribune-Review*, September 7, 2005.

226. Baum, "Deluged"; D. Barry and J. Longman, "A Police Department Racked by Doubt and Accusations," *The New York Times*, September 30, 2005. http://query.nytimes.com/gst/fullpage.html?res=9D04E3DB1330F933A0575AC0A9639C8B63&sec=&spon= (accessed May 2009); Perlstein and Lee, "The Good and the Bad."

227. Perlstein and Lee, "The Good and the Bad."

228. Baum, "Deluged."

229. Arey and Wilder, "NOPD SWAT."

230. Perlstein and Lee, "The Good and the Bad."

231. C. Connolly, "Katrina's Emotional Damage Lingers," *The Washington Post*, December 7, 2005. http://www.washingtonpost.com/wp-dyn/content/article/2005/12/06/AR2005120601594.html. (accessed May 2009).

232. Perlstein and Lee, "The Good and the Bad"; Associated Press, "Katrina's Devastation."

233. Perlstein and Lee, "The Good and the Bad."

234. Baum, "Deluged."

235. Ibid.

236. Ibid.

237. D. Firestone and R. Pérez-Pe–a, "Failure Reveals Creaky System, Experts Believe," *New York Times*, August 15, 2003, A1, 18.

238. D. McGinn and K. Naughton, "The Price of Darkness," *Newsweek*, 142, no. 8 (August 25, 2003), 42.

239. A. Todd and A. Wood, "'Flex Your Power': Energy Crises and the Shifting Rhetoric of the Grid," *Atlantic Journal of Communication*, 14, no. 4 (2006): 212.

240. P. Bourdieu and L. Wacquant, "The New Planetary Vulgate," in *Pierre Bourdieu Political Interventions: Social Science and Political Action*, selected and introduced F. Poupeau and T. Discepolo (London: Verso, 2008), 367.

241. F. Jameson, *Postmodernism, or the Cultural Logic of Late Capitalism* (Durham, NC: Duke University Press, 1991), ix.

242. B. Latour, *We Have Never Been Modern* (London: Harvester Wheatsheaf, 1993), 13.

243. Ibid., 34.

244. Ibid., 4.

245. T. Luke, "The Coexistence of Cyborgs, Humachines, and Environments in Postmodernity: Getting Over the End of Nature," in *The Cybercities Reader*, ed. S. Graham, 106–10 (Oxford, UK: Blackwell, 2003).

246. B. Latour, *We Have Never Been Modern* (London: Harvester Wheatsheaf, 1993), 34.

247. U.S.-Canada Power Outage Task Force, *Final Report on the August 14, 2003 Blackout in the United States and Canada: Causes and Recommendations* (April 2004), 5. https://reports.energy.gov/BlackoutFinal-Web.pdf (accessed May 16, 2008).

248. Latour, *We Have Never Been Modern*, 10–11.

249. Ibid., 11.

250. Ibid.

251. M. Castells, *The Informational City* (Oxford, UK: Blackwell, 1989).

252. See S. Giedion, *Mechanization Takes Command: A Contribution to Anonymous History* (New York: Norton, 1948); Anson Rabinbach, *The Human Motor: Energy, Fatigue, and the Origins of Modernity* (New York: Basic, 1990); and C. Pursell, *The Machine in America: A Social History of Technology* (Baltimore: Johns Hopkins University Press, 1995).

253. M. Brown, B. Sovacool, and R. Hirsh, "Assessing U.S. Energy Policy," *Daedalus* (Summer 2006), 7.

254. See D. Nye, *The Technological Sublime* (Cambridge, MA: MIT Press, 1996); L. Mumford, *Technics and Civilization* (New York: Harcourt Brace Jovanovich, 1963).

255. See D. Nye, *Electrifying America: Social Meanings of a New Technology* (Cambridge, MA: MIT Press, 1990).

256. See M. Foucault, *Technologies of the Self* (Amherst: University of Massachusetts Press, 1988).

257. M. Foucault, *The Foucault Effect: Studies in Governmentality*, G. Burchell, C. Gordon, and P. Miller, eds. (Chicago: University of Chicago Press, 1991), 87–104.

258. T. Luke, "Real Interdependence: Discursivity and Concursivity in Global Politics," in *Language, Agency, and Politics in a Constructed World*, ed. F. Debrix (Armonk, NY: M. E. Sharpe, 2003), 101–20.

259. C. Perrow, *Normal Accidents: Living with High Risk Technologies* (New York: Basic Books, 1984), 3.

260. Ibid.

261. S. Graham, "Switching Cities Off: Urban Infrastructure and U.S. Air Power," *City*, 9, no. 2 (July 2005): 170–94.

262. J. Adler, "The Day the Lights Went Out," *Newsweek*, 142, no. 8, August 25, 2003, 50.

263. J. Glanz, "Its Coils Tighten, and the Grid Bites Back," *New York Times*, August 17, 2003, 4.

264. D. Yergin and L. Makovich, "The System Did Not Fail, Yet the System Did," *New York Times*, August 17, 2003, 4–10.

265. Glanz, "Its Coils Tighten," 4–5.

266. See R. Hirsh, *Power Loss: The Origins of Deregulation and Restructuring in the American Electric Utility System* (Cambridge, MA: MIT Press, 1999).

267. N. Banerjee and D. Firestone, "New Kind of Electricity Market Strains Old Wires Beyond Limits," *New York Times*, August 24, 2003, A1, 26.

268. B. Stone, 2003, "How to Fix the Grid," *Newsweek,* 142, no. 8, August 25, 2003, 38.

269. G. Anderson et al., "Causes of the 2003 Major Grid Blackouts in North America and Europe, and Recommended Means to Improve System Dynamic Performance," *IEEE Transactions of Power Systems*, 20, no. 4 (November, 2005): 1922–28.

270. M. Wald, R. Pérez-Pe–a, and N. Banerjee, "Experts Asking Why Problems Spread So Far," *New York Times*, August 16, 2003, A1, B4.

271. Ibid.

272. R. Pérez-Pe–a, and M. Wald, "Basic Failures by Ohio Utility Set Off Blackout, Report Finds," *New York Times,* November 20, 2003, A1, 28.

273. See North American Electric Reliability Council (NAERC), *August 14, 2003 Blackout: NAERC Actions to Prevent and Mitigate the Impacts of Future Cascading Blackouts* (Princeton, NJ: NAERC, 2004).

274. R. Pérez-Pe–a, "Utility Could Have Halted '03 Blackout, Panel Says," *New York Times,* April 6, 2004, A16.

275. Associated Press, "Blackout Inquiry Result Could Come by End of Week," *Roanoke Times,* August 17, 2003, A1, 9.

276. McGinn and Naughton, *Price of Darkness,* 42.

277. See NAERC, *August 14, 2003.*

278. See P. Virilio, *A Landscape of Events* (Cambridge: MA: MIT Press, 2000); and, P. Virilio, *The Art of the Motor* (Minneapolis: University of Minnesota Press, 1995).

279. K. Marx and F. Engels, "The Communist Manifesto," in *The Marx-Engels Reader*, ed. Robert C. Tucker (New York: Norton, 1978), 476.

280. See A. Smith, *The Wealth of Nations* (London: Penguin, 1987); T. Luke, "Identity, Meaning and Globalization: Space–Time Compression and the Political Economy of Everyday Life," in *Detraditionalization: Critical Reflections on Authority and Identity*, ed., S. Lash, P. Heelas, and P. Morris (Oxford, UK: Blackwell, 1996), 109–133; and, J. Baudrillard, *The System of Objects* (London: Verso, 1996).

281. See U. Beck, *Risk Society* (London: Sage, 1992).

282. See T. Luke, *Capitalism, Democracy, and Ecology: Departing from Marx* (Urbana: University of Illinois Press, 1999).

283. J. Baudrillard, *For a Critique of the Political Economy of the Sign* (St. Louis: Telos Press, 1981), 85.

284. Ibid., 82. See also B. Agger, *Fast Capitalism* (Urbana: University of Illinois Press, 1989).

285. See P. Bourdieu, *Distinction: A Social Critique of the Judgment of Taste* (Cambridge, MA: Harvard University Press, 1984).

286. Graham, "Switching Cities Off," 170–194.

287. M. Foucault, *History of Sexuality* (New York: Vintage, 1980), 1: 98.

288. T. Luke, *Screens of Power: Ideology, Domination and Resistance in Informational Society* (Urbana: University of Illinois Press, 1989).

289. Foucault, *History*, 106.

290. Ibid.

291. Ibid., 107.

292. See D. Harvey, *A Brief History of Neoliberalism* (Oxford, UK: Oxford University Press, 2005).

293. M. Foucault, *History*, 94.

294. See L. Greenfeld, *The Spirit of Capitalism: Nationalism and Economic Growth* (Cambridge, MA: Harvard University Press, 2001).

295. M. Foucault, *Language, Counter-Memory, Practice: Selected Essays and Interviews*, ed. D. Bouchard (Ithaca, NY: Cornell University Press, 1977), 146.

296. Smith, *Wealth of Nations*, 105.

297. See R. Pool, *Beyond Engineering: How Society Shapes Technology* (Oxford, UK: Oxford University Press, 1997).

298. See Luke, *Capitalism, Democracy and Ecology*, 217–46.

299. See Virilio, *Art of the Motor*; and, Thomas F. Tierney, *The Value of Convenience: A Genealogy of Technical Culture* (Albany: SUNY Press, 1993).

300. See Glanz, "Its Coils Tighten," 4–1.

301. Ibid.

302. Beck, *Risk Society*, 222.

303. Ibid., 233.

304. Foucault, *Governmentality,* 92–96.

305. See Hirsh and Klaidman, "What Went Wrong," *Newsweek*, 142, no. 8, August 25, 2003, 33–39.

306. Luke, *Screens of Power*, 207–39.

307. B. Agger, *Fast Capitalism,* 82.

308. K. Varnelis, "The Centripetal City."

309. Already comprising over 10 percent of the nonoil revenue in the UAE, and with global revenue accounting for $3.4 trillion in 2006 and projected growth at 4.5 percent annually for the next five years, logistics offers a promising risk, see *UAE Yearbook* Economic Development," 2007, http://www.uaeinteract.org/uaeint_misc/pdf_2007/English_2007/eyb5.pdf (accessed February 12, 2008).

310. Planetizen: The Planning and Development Network, "Coming Soon: The World's Largest Airport and 'Logistics City,'" http://www.planetizen.com/node/19718 (accessed April 2, 2008).

311. Arabian Business, "Dubai Opens First 'Luxury' Labour Camp," 2007, http://www.arabianbusiness.com/index.php?option=com_content&view=article&id=8473 (accessed April 11, 2008).

312. See J. DeParle, "Fearful of Restive Foreign Labor, Dubai Eyes Reforms," *New York Times*, August 6, 2007, http://www.nytimes.com/2007/08/06/world/middleeast/06dubai.html?_r=1 (accessed May 8, 2009): "Beds" best describes the extent of the facilities, as they are designed according to a bare minimum provision of space. New regulations developed in response to the workers' protests about inhumane living conditions provide that there should be no more than eight people to a room and toilet, and each person should have a minimum of 3 meters floor space.

313. Human Rights Watch, "Building Towers, Cheating Workers: Exploitation of Migrant Construction Workers in the United Arab Emirates," 2006, http://www.hrw.org/en/reports/2006/11/11/building-towers-cheating-workers (accessed March 4, 2008).

314. J. Shaoul, "The Plight of the UAE's Migrant Workers: The Flipside of a Booming Economy," November 9, 2007, http://www.wsws.org/articles/2007/nov2007/duba-n09.shtml (accessed April 1, 2008).

315. H. Fattah and E. Lipton, "Gaps in Security Stretch from Model Port in Dubai to U.S.," *New York Times*, February 26, 2006, http://www.nytimes.com/2006/02/26/national/26port.html (accessed May 2009).

316. S. Flynn, "The DP World Controversy and the Ongoing Vulnerability of U.S. Seaports," Testimony before the House Armed Services Committee. March 2, 2006. http://www.cfr.org/publication/9998/ (accessed January 13 2008); J. A. Cooke, "The DP World Controversy Could Help our Seaports," *Logistics Management* 4 (2006): 1.

317. M. Foucault, *Security, Territory, Population: Lectures at the College de France* 1977–1978 (Basingstoke, UK: Palgrave Macmillan, 2007), 12.

318. Foucault, *Security,* 20.

319. S. Elden, "Governmentality, Calculation, Territory," *Environment and Planning D: Society and Space* 25 (2007): 564.

320. H. Lefebvre, *The Production of Space* (Oxford, UK: Blackwell, 1981), 41.

321. M. Huxley, "Spatial Rationalities: Order, Environment, Evolution and Government, *Social and Cultural Geography* 7, no. 5 (2006): 773–74.

322. S. Elden, *Mapping the Present: Heidegger, Foucault and the Project of a Spatial History* (London: Continuum, 2001), 145.

323. Christine Boyer, *Dreaming the Rational City: The Myth of American City Planning*. (Cambridge, MA: MIT Press, 1983), 9.

324. Ibid., 33.

325. M. Cooper, "Infrastructure and Event: Urbanism and the Accidents of Finance." (Paper presented to CUNY Center for Place, Culture and Politics, New York, 2008).

326. Foucault, Security, 13.

327. Organization for Economic Cooperation and Development (OECD), *Security in Maritime Transport: Risk Factors and Economic Impact,* Maritime Transport Committee Report (2003): 2 at http://www.oecd.org/LongAbstract/0,3425,en_2649_34367_4429374_119666_1_1_1,00.html (accessed May 2009).

328. Department of Homeland Security "Strategy to Enhance International Supply Chain Security," 2007, http://www.dhs.gov/xprevprot/publications/gc_1184857664313.shtm (accessed July 27, 2007).

329. Quoted in "When Trade and Security Clash," *The Economist*, April 4, 2002, http://www.economist.com/displaystory.cfm?story_id=1066906 (accessed May 2009).

330. Homeland Security "Container Security Initiative," 2005, http://www.cbp.gov/xp/cgov/border_security/international_activities/csi/csi_in_brief.xml (accessed December 12,2007).

331. Customs and Border Patrol (CBP), "Container Security Initiative, 2006–2011 Strategic Plan," 2006, http://www.customs.treas.gov/linkhandler/cgov/border_security/international_activities/csi/csi_strategic_plan.ctt/csi_strategic_plan.pdf (accessed July 2, 2007).

332. For an extended discussion of the TWIC program, please see Deborah Cowen, "A Geography of Logistics: Reterritorializing Economy, Security and the Social." *Annals of the Association of American Geographers* (forthcoming).

333. J. Haveman and H. Shatz, eds., "Protecting the Nation's Seaports: Balancing Security and Cost," Public Policy Institute of California, 2006, http://faculty.spa.ucla.edu/zegart/pdf/R_606JHR.pdf (accessed September 1, 2007).

334. See P. Kraska, *Militarizing the American Criminal Justice System: The Changing Roles of the Armed Forces and the Police* (Boston: Northeastern University Press, 2001); E. Hobsbawm, "The Future of War and Peace," *CounterPunch*, February 27, 2002. http://www.counterpunch.org/hobsbawm1.html; P. Andreas and R. Price, "From War Fighting to Crime Fighting: Transforming the American National Security State," *International Studies Association* 3, no. 3 (2001): 31–52; D. Campbell, "The Biopolitics of Security: Oil, Empire, and the Sports Utility Vehicle," *American Quarterly* 57, no. 3 (2005): 943–72; D. Cowen and N. Smith "After Geopolitics? From the Geopolitical Social to Geoeconomics." *Antipode,* 41 no. 1, 22–48.

335. S. Collier and A. Lakoff, "On Vital Systems Security." (Berkeley, California: Anthropology of the Contemporary Research Collaboratory, 2006), http://www.anthropos-lab.net (accessed March 29, 2007).

336. S. Collier and A. Lakoff, "How Infrastructure Became a Security Problem," in *The Changing Logics of Risk and Security*, ed. Myriam Dunn (New York: Routledge, forthcoming).

337. See D. Cowen, "National Soldiers and the War on Cities," *Theory & Event* 10, no.2 (2007) at http://muse.jhu.edu/login?uri=/journals/theory_and_event/v010/10.2cowen.html (accessed May 2009); for an analysis that focuses on the political dimensions of the emergence of social insurance and social security, Donzelot (1988) provides an account that emphasizes the key role of class struggles over workplace injuries in prompting the development of these techniques. I have argued that it was not only the factory workplace that provoked the experimentation with political technologies that yielded social insurance, but the military that was an important site (see Cowen 2007).

338. K. Rygiel, "The Securitized Citizen," in *Recasting the Social in Citizenship*, ed. E. Isin (Toronto: University of Toronto Press, 2008).

339. Congressional Research Service, "Border and Transportation Security: The Complexity of the Challenge," 2005, http://www.fas.org/sgp/crs/homesec/RL32839.pdf (accessed 29 March 2007).

4; The White House, The National Strategy for Homeland Security, 2002, http://www.whitehouse. gov/homeland/book/nat_strat_hls.pdf (accessed March 29, 2007), 22.

340. See RAND "Evaluating the Security of the Global Containerized Supply Chain," Infrastructure, Safety and Environment, Technical Report, 2004. http://www.rand.org (accessed March 21, 2007).

341. Lieutenant Colonel Thomas Goss is an active duty officer in the U.S. Army serving on the International Military Staff at NATO Headquarters. Goss earned a PhD in history from Ohio State University and a master's degree in homeland security from the Naval Postgraduate School.

342. N. Rose, *Governing the Soul: The Shaping of the Private Self*, 2nd ed. (London: Free Association Books, 1999), 141.

343. See Cooper, "Infrastructure," 2008; Graham and Marvin, *Splintering Urbanism*.

344. See H. Jomini, *The Art of War*, trans. G. H. Mendell (Philadelphia: J.D. Lippincott, 1862).

345. M. DeLanda, *War in the Age of Intelligent Machines* (Brooklyn, NY: Zone Books, 1991).

346. E. Smykay and B. LaLonds, *Physical Distribution: The New and Profitable Science of Business Logistics* (Chicago: Dartnell Press, 1967).

347. B. Allen, "The Logistics Revolution and Transportation," *Annals of the American Academy of Political and Social Science*, 553 (1997): 114.

348. G. Davis, *Logistics Management* (Lexington, MA: D.C. Heath, 1974).

349. See D. Cowen, "A Geography of Logistics: Reterritorializing Economy, Security and the Social," *Annals of the Association of American Geographers* (forthcoming).

350. L. Busch, "Performing the Economy, Performing Science: From Neoclassical to Supply Chain Models in the Agrifood Sector," *Economy and Society* 3 (2007): 441.

351. E. Bonacich, "Labor and the Global Logistics Revolution," in Richard Appelbaum, William Robinson (Eds.) *Critical Globalization Studies* (New York: Routledge, 2005), 357–77.

352. A. Barry, "The Anti-Political Economy," *Economy and Society* 31, no. 2 (2002): 268–84.

353. Ibid.

354. M. Callon, ed., *The Laws of the Markets* (London: Blackwell,1998).

355. W. Brown, "American Nightmare: Neoliberalism, Neoconservatism, and De-Democratization," *Political Theory* 34, no. 6 (2006): 693.

356. D. Sui, "Musing on the Fat City: Are Obesity and Urban Forms Linked?" *Urban Geography* 24 (2003), 75–84.

357. R. Park, E. Burgess, and R. McKenzie, *The City* (Chicago: Chicago University Press, 1967).

358. D. Harvey, "The City as Body Politic," in *Wounded Cities: Destruction and Reconstruction in a Globalized World*, ed. Jane Schneider and Ida Susser (Oxford, UK: Berg, 2003), 34.

359. E. Shell, *Fat Wars: The Inside Story of the Obesity Industry* (London: Atlantic, 2002).

360. See WHO, *Obesity: Preventing and Managing the Global Epidemic: Interim Report of a WHO Consultation on Obesity* (Geneva, Switzerland: World Health Organization, 1997).

361. See the International Obesity Task Force Web site, http://www.iotf.org.

362. See J. Revill, "A Deadly Slice of American Pie," *The Observer,* September 21, 2003, 17. http:// www.guardian.co.uk/society/2003/sep/21/health.lifeandhealth.

363. See *Men's Fitness* Web site, "Survey Methodology." http://www.mensfitness.com/rankings/3.

364. See D. Reynolds. "Your City's Fat—Now What? Being America's Fattest City Isn't Good For One's Image," *ABC News*, January 4, 2002. http://abcnews.go.com/sections/wnt/worldnewsTonight/ fatcity020104.html.

365. See Get Lean Houston. http://www.getleanhouston.com.

366. See *Men's Fitness*. "Houston." http://www.mensfitness.com/rankings/221.

367. Sui, "Musing on the Fat City."

368. See R. Killingsworth, J. Earp, and R. Moore. "Introduction: Supporting Health Through Design: Challenges and Opportunities," *American Journal of Health Promotion* September/October (2003): 1–3.

369. See R. Jackson, "The Impact of the Built Environment on Health: An Emerging Field," *American Journal of Public Health* 93 (2003): 1382–84.

370. See K. Fackelmann. "Studies Tie Urban Sprawl To Health Risks, Road Danger: Driving Every Day Adds Up on Americans' Waistline, and Can Be Dangerous For Their Health," *USA Today*, August 29, 2003, http://www.usatoday.com/news/nation/2003-08-28-sprawl-usat x.html.

371. Sui, "Musing on the Fat City," 82.

372. See Graham and Marvin, *Splintering Urbanism.*

373. See Graham and Marvin, *Splintering Urbanism;* and Jane Summerton, ed., *Changing Large Technical Systems* (Boulder, CO: Westview Press, 1994).

374. M. Gandy, "The Paris Sewers and the Rationalization of Urban Space," *Transactions of the Institute of British Geographers* 24 (1999): 24.

375. See R. Southerland, "Sewer Fitness: Cutting the Fat," *American City and County,* October 1, 2002, http://americancityandcounty.com/mag/government_sewer_fitness_cutting/index.html.

376. B. Newman, "The Sewer-Fat Crisis Stirs a National Stink: Sleuths Probe Flushing," *The Wall Street Journal* June 4, 2001. http://www.viridiandesign.org/notes/251-300/00257_sewer_fat_crisis.html.

377. Ibid.

378. Southerland "Sewer Fitness."

379. See National Association of Sewer Service Companies, http://www.nassco.org.

380. See V. Pagano, "Letter to Nitin Patel, RE: Zircon Industries, Inc.," June 30, 1999, Administrative Supervisor, NYC Department of Environmental, Protection Bureau of Water and Sewer Operations. http://www.greenchem.com/cityofnewyork.html.

381. See M. Allen, "Company Demonstrates Product in City," *The New Bedford Standard Times,* December 22, 2002. http://www.bactapur.com/english/pages/bacta710.html.

382. See Southerland "Sewer Fitness."

383. See City of New York. "`Waste Water Treatment: Preventing Discharges into Sewers: Guidelines for New York City Businesses," Department of Environmental Protection, The City of New York, 2002, http://www.nyc.gov/html/dep/html/grease.html.

384. See Water Environment Federation, "Fat-Free Sewers: How to Prevent Fats, Oils, and Greases From Damaging Your Home and the Environment," 1999, http://www.wef.org/publicinfo/FactSheets/ fatfree.jhtml.

385. See Sfgate, "Southbound 101 Fully Reopens After Grease Spill," Sfgate, January 7, 2004. http://www.sfgate.com.

386. See Environmental Protection Agency, "Biodiesel: Fat to Fuel," http://epa.gov/region09/waste/biodiesel/index.html.

387. Environmental Protection Agency, "From Fat to Fuel; Santa Cruz Shares Results from First-In-Nation Community-Based Biodiesel Production," *Press Release* 28 A, April 28, 2008, http://yosemite.epa.gov/OPA/ADMPRESS.NSF/d0cf6618525a9efb85257359003fb69d/f012d2ab62a3bdde85257439006731ba!OpenDocument.

388. U.S. Environmental Protection Agency, "Biodiesel: Fat to Fuel."

389. U.S. Environmental Protection Agency, "From Fat to Fuel."

390. U.S. Environmental Protection Agency, "Biodiesel: Fat to Fuel."

391. U.S. Environmental Protection Agency, "From Fat to Fuel."

392. See M. Engel, "Land of the Fat," *The Guardian*, May 2, 2002, http://www.guardian.co.uk/world/2002/may/02/usa.medicalscience.

393. See Environmental Protection Agency, "From Fat to Fuel."

394. A. Lake and T. Townshend, "Obesogenic Environments: Exploring the Built and Food Environments, *Perspectives in Public Health* 126 (2006): 262–67

395. Earthtimes "Liposuctioned Fat Could Be Bio-Diesel Fuel" *Earthtimes* December 7, 2006, http://www.earthtimes.org/articles/show/12174.html.

396. J. Sugden. "Doctor Used 'Human Fat To Power Car'." *New York Times,* December 24, 2008 http://www.timesonline.co.uk/tol/news/environment/article5393763.ece (accessed May 2009).

397. J. Urry, *Global Complexity* (Cambridge, UK: Polity 2003).

398a. World Health Organization official, September 2005.

398b. John Urry, *Mobilities* (Cambridge, UK: Polity, 2007), 149.

399. M. Callon and J. Law, "Guest Editorial," *Environment and Planning D: Society and Space* 22 (2004): 4.

400. M. Davis, *The Monster at Our Door: The Global Threat of Avian Flu* (New York and London: New Press, 2005), 79.

401. See R. Salehi and S. Ali, "The Social and Political Context of Disease Outbreaks: The Case of SARS in Toronto," *Canadian Public Policy* 23, no. 4 (2006): 573–86. Ali and Keil (Forthcoming) "Public Health and the Political Economy of Scale: Implications for Understanding the Response to the 2003 Severe Acute Respiratory Syndrome (SARS) Outbreak in Toronto," in *Leviathan Undone: Towards a Political Economy of Scale,* ed. R. Keil and R. MahonVancouver: UBC Press); Roger Keil and S. Harris Ali, "Governing the Sick City: Urban Governance in the Age of Emerging Infectious Disease," *Antipode* 40, no. 1 (2007): 846–71.

402. World Health Organisation, "SARS: Chronology of a Serial Killer."(Update 95 2004), http://www.who.int/csr.

403. J. Bowen and C. Laroe, "Airline Networks and the International Diffusion of Severe Acute Respiratory Syndrome (SARS)," *The Geographical Journal* 172, no. 2 (2006): 130–44.

404. S. Ali and R. Keil, "Global Cities and the Spread of Infectious Disease: The Case of Severe Acute Respiratory Syndrome (SARS) in Toronto, Canada," *Urban Studies* 43 (2006):1–19.

405. D. Heymann, M. Kindhauser, and G. Rodier, "Coordinating the Global Response," in *SARS: How a Global Epidemic Was Stopped,* ed. WHO (Geneva: WHO, 2005), 243–54).

406. D. Fidler, "The Globalization of Public Health: The First 100 Years of International Health Diplomacy," *Bulletin of the World Health Organization* (2001): 842–49.

407. Cited in Gostin et al., "Ethical and Legal Challenges Posed by Severe Acute Respiratory Syndrome: Implications for the Control of Severe Infectious Disease Threats," *Journal of the American Medical Association* 290 (2003): 3229–37.

408. Bowen and Laroe, "Airline Networks and International Diffusion," 130–44.

409. Interview by the authors: November 11, 2005.

410. NACSPH, Learning from SARS: Renewal of Public Health in Canada (Ottawa: Health Canada, 2003). See also, David Walker, *For the Public's Health: Initial Report of the Ontario Expert Panel on SARS and Infectious Disease Control* (Toronto: Ministry of Health and Long-Term Care, 2004).

411. G. Galloway, "WHO Concedes Advisory Damaged Toronto," *The Globe and Mail*, May 1, 2003, P1–2.

412. NACSPH, Learning from SARS.

413. Ibid.

414. Urry, *Mobilities.*

415. Graham, "Constructing Premium Network Spaces: Reflections on Infrastructure Networks and Contemporary Urban Development," *International Journal of Urban and Regional Research* 24 (2000): 183–200.

416. Doganis, *The Airport Business* (New York: Routledge, 1992).

417. Ibid., 443.

418. P. Adey, "Airports, Mobility and the Calculative Architecture of Affective Control," *Geoforum* 39 (2008): 445.

419. M. Gottdiener, *Life in the Air* (Oxford, UK: Rowman and Littlefield, 2004), 187.

420. M. Auge, *Non Places* (London: Verso, 1995), 94. For a critique of this conception of the airport as a "nonplace," see Urry (*Mobilities,* 147) who argues that airport spaces have considerable social

complexity and that the nonplace conception is based on an inaccurate sedentarist notion of place as unchanging and given.

421. M. Callon and J. Law, "Guest Editorial," *Environment and Planning D: Society and Space* 22 (2004): 3–11.

422. W. Voigt, "From the Hippodrome to the Aerodrome, From the Air Station to the Terminal: European Airports 1909–1945," in *Building for Air Travel: Architecture and Design for Commercial Aviation*, ed. J. Zukowsky (Chicago: Prestel, 1996), 42.

423. H. Van Houtum and T. van Naerssen "Bordering, Ordering and Othering." *Tijdschrift Voor Economische En Sociale Geografie* 93 (2002): 125–36.

424. Adey, "Airports, Mobility and the Calculative Architecture," 438.

425. Urry, *Mobilities*, 42.

426. M. Foucault, *The Birth of the Clinic* (London: Tavistock, 1976).

427. Urry, *Mobilities*, 142.

428. D. L. Rhoades, *Evolution of International Aviation* (Burlington VT: Ashgate 2003). See also, B. Srivastava, *Aviation Terrorism* (New Delhi: Manas, 2002) and K. Sweet, *Aviation and Airport Security: Terrorism and Safety Concern* (Upper Saddle River, NJ: Pearson Prentice Hall, 2004).

429. Urry, *Mobilities*, 149.

430. M. Dodge and R. Kitchin, "Flying Through Code/Space: The Real Virtuality of Air Travel," *Environment and Planning A* 36(2004): 195–211.

431. Urry, *Mobilities,* 144.

432. Urry, *Mobilities,* citing Dillon, "Virtual Security: A Life Science of (Dis)order," *Millennium* 32 (2003): 531–58.

433. C. Bennett, "What Happens When You Book An Airline Ticket (Revisited): The Computer Assisted Passenger Profiling System and the Globalization of Personal Data," in *Global Surveillance and Policing: Borders, Security, Identity*, ed. E. Zureik and M. Salter (Uffculme, Devon, UK: Willan, 2006), 113–38; M. Curry, "The Profiler's Question and the Treacherous Traveller: Narratives of Belonging in Commercial Aviation," *Surveillance and Society* 1, no.4 (2004): 475–499.

434. NACSPH, National Report on SARS.

435. P. Sarasin, *Anthrax: Bioterror as Fact and Fantasy* (Cambridge, MA: Harvard University Press, 2006).

436. M. Serres, *Angels: A Modern Myth* (New York: Flammarion, 1995),19, cited by Urry, *Mobilities*, 138.

437. Urry, *Mobilities*, 142.

438. Ibid.

439. B. Latour, *Pandora's Hope: Essays on the Reality of Science Studies* (Cambridge, MA: Harvard University Press, 1999), 193.

440. Interview, WHO Medical Officer, Community Disease Surveillance and Response, September 27, 2005.

441. Interview, GTAA official, December 16, 2005.

442. Ibid.

443. Interview, January 9, 2006.

444. V. Goel, "What Do We Do with the SARS Reports," *Healthcare Quarterly* 7, no.3 (2004): 28–41.

445. Ibid.

446. R. Gottschalk, "Katastrophenschutz an internationalen Flughäfen," Contribution to Public Health Forum (Public Health bei Katastrophenfällen, 2006); personal communication from author.

447. Interview, GTAA official, December 16, 2005.

448. Ibid.

449. Ibid.

450. Graham and Marvin, Splintering Urbanism.

451. Interview, GTAA official, December 16, 2005.

452. W. Gaber and R. Gottschalk, "Entry- and Exit-Screening Procedures at Frankfurt International Airport" (referring to the IHR/WHO 2005)" (Presentation, Frankfurt, May 20, 2007); personal communication from the authors.

453. http://www.gtaa.com/en/gtaa_corporate/ accessed June 2008.

454. Interview November 25, 2005.

455. Ibid.

456. Graham and Marvin, *Splintering Urbanism*.

457. R. Smith, "World City Topologies," *Progress in Human Geography* 27 (2003): 561–82.

458. Ali and Keil, "Contagious Cities," 1207–26.

459. Keil and Ali, "Governing the Sick City," 846–71; and also see, Roger Keil, and Harris Ali, "Multiculturalism, Racism and Infectious Disease in the Global City," *Topia: Canadian Journal of Cultural Studies* 16 (2006): 23–49.

460. R. Gottschalk and W. Preiser, "Bioterrorism: Is It a Real Threat? *Medical Microbiology and Immunology* 194 (2005): 13.

461. E. Wagner, "The Practice of Biosecurity in Canada: Public Health Legal Preparedness and Toronto's SARS Crisis," *Environment and Planning A* 40 (2008): 1647–63.

462. D. Harvey, "Globalization and the Body," in *Possible Urban Worlds,* ed. INURA (Basel: Birkhäuser, 1998).

463. S. Flusty, *De-Coca-Colonization: Making the Globe from the Inside Out* (London and New York: Routledge, 2004); R. Keil, *Los Angeles: Globalization, Urbanization, and Social Struggles* (Chichester, UK: Wiley, 1998; S. Sassen, *Globalization and Its Discontents: Essays on the New Mobility of People and Money* (New York: New Press 1998).

464. C. Major, "Affect Work and Infected Bodies: Biosecurity in an Age of Emerging Infectious Disease," *Environment and Planning A* 40, no. 7 (2008): 1633.

465. U. Beck, *The Risk Society: Towards A New Modernity* (London: Sage, 1992).

466. U. Beck, *Weltrisikogesellschaft* (Frankfurt: Suhrkamp.2007), 312–13.

467. See also R. Keil and S. Ali, "The Avian Flu: The Lessons Learned from the 2003 SARS Outbreak in Toronto," *Area* 38, no. 1 (2006): 107–109.

468. R. Gottschalk and W. Preiser, "Bioterrorism: Is It a Real Threat?"113.

469. For a discussion see S. Ali, "SARS as an Emergent Complex: Toward a Networked Approach to Urban Infectious Disease," in *Networked Disease: Emerging Infections in the Global City*, ed. S. Ali and R. Keil (New York: Wiley Blackwell, 2008), 233–49.

470. J. Mayer, "Geography, Ecology and Emerging Infectious Diseases," *Social Science and Medicine* 50 (2000): 937–52.

471. S. Lau, K. Woo, Y. Li, H. Huang, B. Tsoi, S. Wong, K. Leung, B. Chan, and K. Yuen. "Severe Acute Respiratory Syndrome Coronavirus-Like Virus In Chinese Horseshoe Bats," *Proceedings of the National Academy of Sciences of the United States* 102, no. 39 (2005): 14040–45.

472. P. Agre, "Imagining the Next War: Infrastructural Warfare and the Conditions of Democracy," *Radical Urban Theory*, September 14, 2001, http://www.rut.com/911/Phil-Agre.html (accessed February 12, 2004), 1.

473. W. Joy, "Why the Future Doesn't Need Us," 239.

474. T. Luke, "Everyday Technics as Extraordinary Threats: Urban Technostructures and Nonplaces in Terrorist Actions," in *Cities, War and Terrorism*, ed. S. Graham, 120–36 (Oxford, UK: Blackwell, 2004).

475. See M. Davis, *Buda's Wagon*.

476. J. Hinkson, "After the London Bombings," *Arena Journal*, 24 (2005): 145.

477. Luke, "Everyday Technics."

478. Ibid.

479. Hinkson, "After the London Bombings, 146.

480. Luke, "Everyday Technics."

481. Hinkson, "After the London Bombings," 149.

482. L. Ansems de Vries, "(The War On) Terrorism: Destruction, Collapse, Mixture, Re-enforcement, Construction," *Cultural Politics* 42 (2008): 185.

483. See A. Wenger and R. Wollenmann, *Bioterrorism: Confronting a Complex Threat* (Boulder, CO: Lynne Rienner, 2007).

484. See, for example, R. Pineiro, *Cyberterror* (New York: Forge Press, 2003). For an excellent discussion of the way fears of "cyberterror" attacks are manipulated and exaggerated, see F. Debrix, *Tabloid Terror: War, Culture and Geopolitics* (New York: Routledge, 2008), chap. 1.

485. Luke, "Everyday Technics."

486. Ibid.

487. P. Deer, "Introduction: The Ends of War and the Limits of War Culture, *Social Text*, 25 (2006): 3.

488. Both quotes are from D. Gregory, "'In Another Time-Zone, the Bombs Fall Unsafely...' Targets, Civilians and Late Modern War," *Arab World Geographer* 9, no. 2(2006): 92.

489. For a detailed history of the relations between modernization theory and Cold War military strategy, see N. Gilman, *Mandarins of the Future: Modernization Theory in Cold War America* (Baltimore: Johns Hopkins University Press, 2003).

490. A. Wright, "Structural Engineers Guide Infrastructure Bombing," *Engineering News Record*, April 3, 2003, http://enr.ecnext.com/coms2/summary_0271-8877_ITM.

491. W. Rostow, *The Stages of Economic Growth: A Non-Communist Manifesto* (Cambridge, UK: Cambridge University Press, 1960).

492. D. Milne "'Our Equivalent of Guerrilla Warfare': Walt Rostow and the Bombing of North Vietnam, 1961–1968," *The Journal of Military History* 71 (January 2007): 169–203.

493. Gilman, *Mandarins of the Future*, 199.

494. As is generally the case, of course, the increasingly intense aerial annihilation only strengthened the resolve of the North Vietnamese civilians, so bolstering Viet Cong power in the process.

495. Cited in Milne "Our Equivalent of Guerrilla Warfare."

496. For an etymology of these phrases, see History News Network, http://hnn.us/articles/30347.html.

497. See Gilman, *Mandarins of the Future*.

498. *New York Times* columnist, T. Friedman, April 23, 1999, cited in I. Skoric, "On Not Killing Civilians," http://www.amsterdam.nettime.org, May 6, 1999 (accessed February 16, 2004).

499. See D. Rumsfeld, News Transcript, U.S. Department of Defense, March 22, 2004, http://www.defenselink.mil/transcripts/transcript.aspx?transcriptid=2361.

500. T. Ansary, An Afghan-American Speaks," "*Salon.Com.* http://archive.salon.com/news/feature/2001/09/14/afghanistan/.

501. Gilman, *Mandarins of the Future*.

502. J. Warden, "The Enemy as a System," *Airpower Journal* 9, no. 1 (1995): 41–55.

503. U.S. Air Force, *Targeting*. Air Force Doctrine Document 2-1.9, Secretary of the Air Force, June 8, 2006, 22–33.

504. P. Barriot and C. Bismuth, "Ambiguous Concepts and, Porous Borders," in *Treating Victims of Weapons of Mass Destruction: Medical, Legal and Strategic Aspects*, ed. Patrick Barriot and Chantal Bismuth (London: Wiley, 2008).

505. The only such weapons that have been made public are the CBU-94 "Blackout Bomb" and the BLU-114/B "Soft-Bomb." http://www.fas.org/man/dod-101/sys/dumb/blu-114.htm.

506. See Federation of American Scientists, CBU-94, "Blackout Bomb" and BLU-114/B "Soft-Bomb," http://www.fas.org/man/dod-101/sys/dumb/blu-114.htm.

507. B. Carroll (2002), "Seeing Cyberspace," http://www.sarai.net/publications/readers/02-the-cities-of-everyday-life/04cyber_ electric.pdf (accessed January 2009).

508. Barriot and Bismuth, "Ambiguous Concepts."

509. Warden, "The Enemy as a System."

510. As the MADRE human rights organization point out, "Attacks on civilians and civilian

infrastructure are grave breaches of international law including: Article 33 of the Fourth Geneva Conventions, Article 48 of the Protocol 1 Additional to the Geneva Conventions, and Article 50 of the Hague Convention." Further, the Rome Statute of the International Criminal Court (ICC) includes as war crimes: "Intentionally directing attacks against the civilian population as such or against individual civilians not taking direct part in hostilities," and "Intentionally directing attacks against civilian objects" (Article 8 2 (b) (i) and (ii))"; see MADRE, "War on Civilians: A MADRE Guide to the Middle East Crisis," July 19, 2006, http://www.madre.org/articles/me/waroncivilians.html.

511. K. Rizer, "Bombing Dual-Use Targets : Legal, Ethical, and Doctrinal Perspectives," *Air and Space Power Chronicles*, January 5, at http://www.airpower.maxwell.af.mil/airchronicles/cc/Rizer.html (accessed February 2001), 1.

512. Ibid.

513. Ibid.

514. Ibid.

515. Ibid.

516. Ibid.

517. E. Felker, *Airpower, Chaos and Infrastructure : Lords of the Rings* Paper No. 14 (Maxwell Air Force Base, Alabama: U.S. Air War College Air University Press,1998): 1–20.

518. Ibid.

519. Source: Redrawn from Edward Felker, *Airpower, Chaos and Infrastructure: Lords of the Rings* (Maxwell Air Force Base, Alabama: U.S. Air War College Air University. Maxwell paper 14 (1998): 12.

520. R. Blakeley, *Bomb Now, Die Later.* Bristol University, Department of Politics, 2003, http://www.geocities.com/ruth_blakeley/bombnowdielater.htm (accessed February 2004), 25.

521. E. Hinman, "The Politics of Coercion," *Toward a Theory of Coercive Airpower for Post–Cold War Conflict* CADRE Paper No. 14. (Maxwell Air Force Base, Alabama: Air University Press, 36112-6615, August 2002), 11.

522. Cited in C. Rowat, "Iraq: Potential Consequences of War," *Campaign Against Sanctions in Iraq Discussion List*, 8 (November 2003): http://www.casi.org.uk/discuss/2002/msg02025.html (accessed February 12, 2004).

523. Cited in Rowat, "Iraq Potential Consequences of War."

524. Rowat, "Iraq Potential."

525. C. Bolkcom and J. Pike, *Attack Aircraft Proliferation: Issues for Concern* (Federation of American Scientists, 1993). http://www.fas.org/spp/aircraft (accessed February 25, 2004), 2.

526. T.. Keaney and E. Cohen, *Gulf War Air Power Surveys* (GWAPS) (Washington, DC ; Johns Hopkins University and the US Air Force, 1993), Vol. 2, Part 2, chap. 6, 23 note 53. http://www.au.af.mil/au/awcgate/awc-hist.htm+gulf (accessed February 15, 2004).

527. Cited in Blakeley, *Bomb Now, Die Later.*

528. R. Blakeley, "Targeting Water Treatment Facilities" (Campaign Against Sanctions in Iraq, Discussion List, January 24, 2003), 2, http://www.casi.org.uk/discuss/2003/msg00256.html.

529. United Nations Children's Fund (UNICEF), *Annex II of S/1999/356, Section 18.* 1999. at http://www.un.org/Depts/oip/reports, February 17, 2004.

530. T. Smith., "The New Law of War : Legitimizing Hi-Tech and Infrastructural Violence," *International Studies Quarterly*, 46 (2002): 365.

531. Cited in Blakeley, *Bomb Now, Die Later.*

532. See chapter 8.

533. See E. Weizmann, *Hollow Land* (London: Verso, 2006).

534. See S. Graham, "Lessons in Urbicide," *New Left Review* 19 (January–February, 2003): 63–73.

535. "Israeli Official Calls For Striking Palestinian Infrastructure," *Arabic News,* May 6, 2001; Rita Giacaman and Abdullatif Husseini, "Life and Health During the Israeli Invasion of the West Bank: The Town of Jenin," *Indymedia Israel* (May 22, 2002).

536. M. Zeitoun, "IDF Infrastructure Destruction by Bulldozer," *Electronic Intifada* (August 2, 2002):

537. Y. Groll-Yaari and H. Assa, *Diffused Warfare: The Concept of Virtual Mass* Haifa, Israel: February 2007, 23.

538. The "Strangling Gaza" subtitle, and the quote, are drawn from C. Chelala, "Strangling Gaza," *Common Dreams*, December 15, 2007, http://www.commondreams.org/archive/2007/12/14/5844/.

539. Gaza is an extended and densely populated city strip roughly the size, at 25 miles long by 6 wide, of England's Isle of Wight. As of 2006, the Strip was inhabited by 1.4 million people; 840,000 of these were children. The density of the population in Gaza is one of the highest in the world. For example, in the Jabalya refugee camp, there are approximately 28,571 people per square mile. See D. Li, "The Gaza Strip as Laboratory: Notes in the Wake of Disengagement," *Journal of Palestine Studies* 35, no. 2 (2006): 40. http://scienceforpeace.sa.utoronto.ca/Gaza_Page/Gaza_Page.html.

540. As Li puts it, "'Closure' is a broad term that includes various restrictions on the circulation of people and goods, ranging from prohibition on international travel to mass house arrest ('curfew')." D. Li, "The Gaza Strip as Laboratory: Notes in the Wake of Disengagement," *Journal of Palestine Studies* 35, no. 2 (2006), 40.

541. Ibid.

542. Ibid.

543. Ibid.

544. It has aso emerged that the later Hamas takeover of the strip in June 2007 was their attempt to forestall a U.S.-funded coup attempt by the rival Fatah organization—an attempt to reverse the result of the democratic election. See S. Milne, "To Blame the Victims For This Killing Spree Defies Both Morality and Sense," *The Guardian*, March 5, 2008, http://www.guardian.co.uk/commentisfree/2008/mar/05/israelandthepalestinians.usa.

545. J. Loewenstein, "Notes from the Field: Return to the Ruin that Is Gaza," *Journal of Palestine Studies*, 36, no. 3 (Spring 2007): 23–35.

546. K. Koning-AbuZayd, "This Brutal Siege of Gaza Can Only Breed Violence in Gaza City," *The Guardian*, January 23, 2008, http://www.guardian.co.uk/commentisfree/2008/jan/23/israelandthepalestinians.world.

547. Electronic Intifada, "Israel invades Gaza: 'Operation Summer Rain'," (June 27 2006). http://electronicintifada.net/bytopic/442.shtml.

548. I. Kimber, "What Happened to the Gaza Strip?" IMEMC News, October 13, 2006. http://www.imemc.org/content/view/22059/117/.

549. S. Milne, "To Blame the Victims."

550. Ibid.

551. At the time the UN was feeding 735,000 Gazans, more than half the territory's population.

552. Palestinian Medical Relief Society, "Public Health Disaster in Gaza Strip: Urgent Appeal for Support to Avert Public Health Disaster in the Gaza Strip," June 27, 2007, http://www.pmrs.ps/last/etemplate.php?id=18.

553. D. Gregory, "'In Another Time-Zone, the Bombs Fall Unsafely…' Targets, Civilians and Late Modern War," *Arab World Geographer* 9, no. 2 (2006): 93.

554. Association of Civil Rights in Israel, Letter to Israeli Minister of Defense, n.d. http://www.phr.org.il/phr/files/articlefile_1151937195636.doc.

555. K. Koning-AbuZayd, "This Brutal Siege of Gaza."

556. K. Zaat, Norwegian Refugee Council (NRC)—Norway, "Isolation of Gaza Must End," Reuters, November 29, 2007, http://www.alertnet.org/thenews/fromthefield/nrc/119632132787.htm.

557. I. Kimber, International Middle East Media Center, *News*, October 13, 2006. http://www.imemc.org/content/view/22059/117/.

558. Cited in, "Statement of Concern for the Public Health Situation in Gaza Open Letter, Canadian Health Professionals," July 31, 2006. http://electronicintifada.net/v2/article5357.shtml.

559. Casre International, "Crisis in Gaza." http://www.care-international.org/index.php?option=com_content&task=view&id=97&Itemid=94.

560. Associated Press, "Malnutrition Common for Gaza Kids," April11, 2007, http://www.jpost.com/servlet/Satellite?pagename=JPost%2FJPArticle%2FShowFull&cid=1176152773887.

561. Cited in, "Statement of Concern."

562. Associated Press, "Four Dead, Thousands Evacuated in Gaza Sewage Flood," March 27, 2007. http://www.iht.com/articles/ap/2007/03/27/africa/ME-GEN-Gaza-Sewage-Flood.php.

563. See R. McCarthy (2009), "Hamas Offers $52m Handouts to Help Hardest-hit Gazans," The Guardian, January 26, 2009. http://www.guardian.co.uk/world/2009/jan/26/hamas-payout-gaza-infrastructure.

564. See G. Rattray, *Strategic Warfare in Cyberspace* (Cambridge, MA: MIT Press, 2001).

565. W. Church, "Information Warfare," *International Review of the Red Cross* 837 (2001): 205–16.

566. U.S. Department of Defense, *Joint Vision 2020* (Washington, DC: U.S. Department of Defense, 2000).

567. Cited in Church, "Information Warfare."

568. D. Onley, "U.S. Aims to Make War on Iraq's Networks," Missouri Freedom of Information Center, 2003. http://foi.missouri.edu/terrorbkgd/usaimsmake.html (accessed February 24, 2004).

569. P. Stone, "Space Command Plans For Computer Network Attack Mission," U.S. Department of Defense: Defense Link, January 14, 2003. http://www.defenselink.mil (February 22, 2004).

570. B. Rosenberg, "Cyber Warriors: USAF Cyber Command Grapples with New Frontier Challenges," *C4ISR Journal* (August 1, 2007).

571. W. Astore, "Attention Geeks and Hackers Uncle Sam's Cyber Force Wants You!" *Tom Dispatch*, (June 5, 2008). http://www.tomdispatch.com/post/174940/william_astore_militarizing_your_cyberspace.

572. Cited in W. Astore, "Attention Geeks and Hackers Uncle Sam's Cyber Force Wants You!" *Tom Dispatch*, June 5, 2008, http://www.tomdispatch.com/post/174940/william_astore_militarizing_your_cyberspace.

573. A term used to describe the deluging of electronic networks which such volumes of traffic that they can't function normally. http://www.networkworld.com/news/2007/051707-estonia-recovers-from-massive-denial-of-service.html.

574. See R. Marquand and B. Arnoldy, "China Emerges as Leader In Cyberwarfare," *Christian Science Monitor,* September 14, 2007. http://www.csmonitor.com/2007/0914/p01s01-woap.html; see also Q. Liang and W. Xiangsui, *Unrestricted Warfare* (Panama: Pan American, 2002).

575. B. Johnson, "NATO Says Cyberwarfare Possesses as Great a Threat as a Missile Attack," *The Guardian*, March 6, 2008. http://www.guardian.co.uk/technology/2008/mar/06/hitechcrime.uksecurity.

576. Steven Metz, "The Next Twist of the RMA," *Parameters* (Autumn 2000): 40–53, http://www.carlisle.army.mil/usawc/Parameters/00autumn/metz.htm.

577. Ibid.

578. Cited in C. Smith, "U.S. Wrestles with New Weapons," NewsMax.Com, March 13, 2003, http://www.newsmax.com/archives/articles/2003/3/134712.shtml.

579. Ibid.

580. A. Behnke, "The Re-enchantment of War in Popular Culture," *Millennium:Journal of International Studies* 34, no. 3 (2006): 937.

581. P. Agre, "Imagining the Next War: Infrastructural Warfare and the Conditions of Democracy," *Radical Urban Theory* (September 14, 2001). http://www.rut.com/911/ Phil-Agre.html.

582. D. Murakami Wood and J. Coaffee, 'Security Is Coming Home: Rethinking Scale and Constructing Resilience in the Global Urban Response to Terrorist Risk," *International Relations* 20, no. 4 (2006): 503–517.

583. P. Agre, "Imagining the Next War."

584. Ibid.

585. See, for example, F. Iklé, *Annihilation From Within: The Ultimate Threat to Nations* (New York: Columbia University Press, 2006).

586. See, for example, R. Sennett, "The Age of Anxiety," *Guardian*, October 23, 2004, http://books. guardian.co.uk/review/story/0,1332840,00.html.

587. Agre, "Imagining the Next War."

588. Ibid.

589. See Canadian Consortium on Human Security, "Human Security Cities," http://www.humansecurity-cities.org/page119.ht.

590. P. Agre, "Imagining the Next War."

591. J. Bellflow, "The Indirect Approach,'" *Armed Forces Journal*, http://www.afji.com/2007/01/2371536.

592. T. Ansary, "A War Won't End Terrorism," *San Francisco Chronicle*, October 19, 2002, relisted at http://www.commondreams.org/views02/1019-02.htm.

593. D. Gregory, *The Colonial Present* (Oxford, UK: Blackwell, 2004).

594. See G. Agamben, *Homo Sacer: Sovereign Power and Bare Life* (Stanford, CA: Stanford University Press, 1998.)

595. D. Gregory, "In Another Time-Zone," 923.

596. D. Gregory, "Defiled Cities," *Singapore Journal of Tropical Geography*, 24, no. 3 (2003): 311.

597. B. Diken and C. Laustsen, "Camping as a Contemporary Strategy: From Refugee Camps to Gated Communities" (AMID Working Paper Series, 32, Aalborg University, 2002).

598. D. Gregory, "Defiled Cities," 313.

599. D. Chakrabarty, *Habitations of Modernity: Essays in the Wake of the Subaltern Studies*. (Chicago: University of Chicago Press, 2002); J. Robinson, *Ordinary Cities: Between Modernity and Development* (London: Routledge, 2006).

600. J. Anjaria, "On Street Life and Urban Disasters: Lessons from a 'Third World' City," in *What Is a City? Rethinking the Urban After Hurricane Katrina*, ed. P. Steinberg and R. Shields (Athens, GA: University of Georgia Press, 2008).

601. Seabrook, J., *In the Cities of the South: Scenes from a Developing World* (London:Verso, 1996), 5.

602. M. Davis, *Planet of Slums* (London: Verso,2006); T. Agnotti, "Review Essay: Apocalyptic Anti-Urbanism: Mike Davis and his Planet of Slums," *International Journal of Urban and Regional Change* 30, no. 4 (2006): 961.

603. See, for instance, R. Koolhaas, "Lagos," in *Mutations*, ed., R. Koolhaas et al. (Bordeaux, France: ACTAR/Barcelona: Arc en Rêve Centre d'Architecture, 2000); D. Mitlin and D.Satterthwaite eds., *Empowering Squatter Citizens: Local Government, Civil Society and Urban Poverty Reduction* (London: Earthscan, 2004).

604. Davis, *Planet of the Slums,* .201; and see discussion in Boal 2006; V. Rao, "Slum as Theory," *International Journal of Urban and Regional Research* 30, no.1 (2006): 225–32.

605. W. Rostow, *The Stages of Economic Growth: A Non-Communist Manifesto*. (Cambridge, UK: Cambridge University Press, 1960).

606. M. Gandy, "Planning, Anti-Planning and the Infrastructure Crisis Facing Metropolitan Lagos," *Urban Studies* 43, no. 2 (2006): 390.

607. A. Amin, "Collective Culture and the Urban Public Space." *City*, 12, no. 1 (2008): 22.

608. A. Appadurai, "Spectral Housing and Urban Cleansing: Notes on Millennial Mumbai," *Public Culture* 12, no. 3(2000): 627–51; S. Patel and D. Mitlin, "Sharing Experiences and Changing Lives." *Community Development Journal* 37, no. 2(2002): 125–36; C. McFarlane, "Geographical Imaginations and Spaces of Political Engagement: Examples from the Indian Alliance," *Antipode* 36, no. 5 (2004): 890–916; C. McFarlane, "Transnational Development Networks: Bringing Development and Postcolonial Approaches into Dialogue," *The Geographical Journal* 172, no. 1(2006): 35–49.

609. McFarlane and Rutherford, "Political Infrastructures."

610. Anjaria, *On Street Life*.

611. Davis, *Planet of the Slums*; P. Ellis, "Chaos in the Underground: Spontaneous Collapse in a Tightly

Coupled System," *Journal of Contingencies and Crisis Management*, 6, no. 3 (1998): 137–52; J. Hyndman and M. de Alwis, "Bodies, Shrines, and Roads: Violence, (Im)mobility, and Displacement in Sri Lanka," *Gender, Place and Culture* 11, no. 4 (2004): 535–57; A. Shaban and R. N. Sharma, "Water Consumption Patterns in Domestic Households in Major Cities," *Economic and Political Weekly* (June 9, 2007): 2195.

612. Davis, *Planet of the Slums*, 125.

613. Cited in M. Siemiatycki, "Message in a Metro: Building Urban Rail Infrastructure and Image in Delhi, India," *International Journal of Urban and Regional Research* 30, no.2 (2006), 285.

614. T. Srivastava, "By Bus, It Was a Bumpy Ride," *Times of India* News Network, December 24, 2002, http://www.timesofindia.com.

615. E. Swyngedouw, "Circulations and Metabolisms: (Hybrid) Natures and (Cyborg) Cities," *Science as Culture* 15, no.2 (2006): 105–22.

616. Siemiatycki, "Message in a Metro," 285.

617. Ibid.

618. P. Patel, "Will $500 Billion Close India's Infrastructure Deficit?" World Bank, 2007, http://go.worldbank.org/7URJM6BOM0 (accessed June 2008).

619. Ibid.

620. Graham and Thrift, "Out of Order."

621. Ellis, 1998; Dennis, K. "Time in the Age of Complexity," *Time and Society*, 16, no. 2–3(2007): 139–55.

622. Swyngedouw, "Message in a Metro," 111–12; Vigarello, 1988.

623. Bunnell and Coe (2005: 841); and see K. Easterling, *Enduring Innocence: Global Architecture and Its Political Masquerades* (Cambridge, MA: Massachusetts Institute of Technology, 2005).

624. K. S. Lee, A. Annas, and G. T. Oh, "Costs of Infrastructure Deficiencies for Manufacturing in Nigerian, Indonesian and Thai Cities," *Urban Studies* 36, no. 12 (1999): 2135–49.

625. D. Dasgupta, "In Mumbai, Pay More for Uninterrupted Power," *Times of India*, April 2, 2008, 4.

626. Fabricius, "Resisting Representation: The Informal Geographies of Rio de Janeiro." *Harvard Design Magazine* 28, no. 1–8 (2008): 5.

627. K. Coelho, "Tapping In: Leaky Sovereignties and Engineered (Dis)Order in an Urban Water System." *Sarai Reader: Turbulence* (New Delhi: Sarai, 2006), 499.

628. Gandy, op cit, 383.

629. Adapted from UN-Habitat, *Global Report on Human Settlements* (London: Earthscan, 2003), 113–14.

630. Gandy op cit, 386.

631. UN Habitat, *Global Report*, 113–14.

632. A. Boland, "The Trickle-Down Effect: Ideology and the Development of Premium Water Networks in China's Cities." *International Journal of Urban and Regional Research*, 31, no. 1(2007): 21–40.

633. S. N. Tesh and E. Paes-Machado, 'Sewers, Garbage and Environmentalism in Brazil," *Journal of Environment and Development* 13, no. 1 (2004): 56

634. A. Shaban and R. N. Sharma, "Water Consumption."

635. Shaban and Sharma, "Water Consumption"; P. Chatterjee, *The Politics of the Governed: Reflections on Popular Politics in Most of the World* (Delhi: Permanent Black, 2004).

636. Patel, "Will $500 Billion."

637. A. Gilbert, "Water For All: How to Combine Public Management with Commercial Practice to the Benefit of the Poor?" *Urban Studies*, 44, no. 8 (2007): 1574.

638. M. Muller, "Parish Pump Politics: The Politics of Water Supply in South Africa." *Progress in Development Studies* 7, no. 1 (2007): 33–45.

639. HSRC/DWAF, 2004.

640. Muller, "Parish Pump Politics," 41.

641. I. Giglioli and E. Swyngedouw, "Lets' Drink to the Great Thirst! Water and the Politics of Fractured Technonatures in Sicily," *International Journal of Urban and Regional Research* 32, no. 2 (2008): 392–414.

642. McFarlane and Rutherford, "Political Infrastructures."

643. Giglioli and Swyngedouw, "Let's Drink."

644. T. Moss, "'Cold Spots' of Urban Infrastructure: 'Shrinking' Processes in Eastern Germany and the Modern Infrastructural Ideal," *International Journal of Urban and Regional Research* 32, no. 2 (2008): 252–70.

645. C. Humphrey, "Rethinking Infrastructure: Siberian Cities and the Great Freeze of January 2001." *Wounded Cities: Deconstruction and Reconstruction in a Globalized World*, ed. J. Schneider and I. Susser, (New York: Berg, 2003), 91–110.

646. Star, Ethnography of Infrascture.

647. Ibid., 94.

648. Ibid., 104.

649. Schneider and Susser, *Wounded Cities*.

650. Davis, *Buda's Wagon*.

651. Harvey, "City as Body Politics," 25–26.

652. The following discussion on Mumbai is based on fieldwork conducted in the state and informal settlements in the year following the 2005 monsoon.

653. Anjaria, "On Street Life."

654. Concerned Citizens Commission Report, *Mumbai's Floods* (Mumbai: CCC, 2005); C. Prabhu, "Why Mumbai Choked." *Frontline*, 22, no. 17 (August 13, 2005b), 26; P. Chandrashekhar, *Mumbai* (Mumbai: MEDC, 2005a).

655. Anjaria, "On Street Life."

656. Ibid.

657. Prabhu, "Why Mumbai Choked."

658. *Times of India* (2005), "State Plans Use of Plastic Bags." *Times of India*, August 25, 2005.

659. For example, Bombay First, Vision Mumbai: Transforming Mumbai into a World-Class City (Mumbai: Bombay First-McKinsey, 2003); C. McFarlane, "Governing the Contaminated City: Sanitation in Colonial and Postcolonial Bombay." *International Journal of Urban and Regional Research* 32, no. 2 (2008): 415–35.

660. M. Majumdar, "Monsoon Dampens Mumbai's Reputation," 2005. BBC, http://news.bbc.co.uk/1/hi/world/south_asia/4727371.stm (accessed June 2008).

661. See Gandy, "Planning, Antiplanning" on Lagos.

662. Moss, "Cold Spots."

BIBLIOGRAPHY

Adey, P. "Airports, Mobility and the Calculative Architecture of Affective Control." *Geoforum* 39 (2008): 438–51.

Adler, J. "The Day the Lights Went Out." *Newsweek*, 142, no. 8 (August 25, 2003): 45–51.

Agamben, G. *Homo Sacer: Sovereign Power and Bare Life*. Stanford, CA: Stanford University Press, 1998.

Agger, B. *Fast Capitalism*. Urbana: University of Illinois Press, 1989.

Agre, P. (2001), "Imagining the Next War: Infrastructural Warfare and the Conditions of Democracy." *Radical Urban Theory* (September 14, 2001): 1. http://www.rut.com/911/Phil-Agre.html (accessed February 12, 2004).

Air Force, Secretary of the. Air Force Doctrine Document 2-1.9. June 8, 2006: 22–33.

Ali, S. H. "SARS as an Emergent Complex: Toward a Networked Approach to Urban Infectious Disease." In *Networked Disease: Emerging Infections in the Global City*, edited by H. Ali and R. Keil, 233–49. New York: Riley Blackwell, 2008.

———. R. Keil. "Global Cities and the Spread of Infectious Disease: The Case of Severe Acute Respiratory Syndrome (SARS) in Toronto, Canada." *Urban Studies* 43 (2006): 1–19.

———. "Contagious Cities." *Geography Compass* 1, no.5 (2007): 1207–26.

———. "Public Health and the Political Economy of Scale: Implications for Understanding the Response to the 2003 Severe Acute Respiratory Syndrome (SARS) Outbreak in Toronto." In *Leviathan Undone: Towards a Political Economy of Scale*, edited by Roger Keil and Rianne Mahon. Vancouver: UBC Press. (Forthcoming).

Allen, B. "Environmental Justice and Expert Knowledge in the Wake of a Disaster." *Social Studies of Science* 37, no. 1 (February 2007): 103–10.

Allen, M. "Company Demonstrates Product in City." *The New Bedford Standard Times*. December, 22, 2002. http://www.bactapur.com/english/pages/bacta710.html

Amin, M. "North America's Electricity Infrastructure: Are We Ready For More Perfect Storms?" *IEEE Security and Privacy Magazine* (September/October 2003): 19–25.

Amoore, L. "Biometric Borders: Governing Mobilities in the War on Terror." *Political Geography* 25 (2006): 336–51.

———. Marieke de Goede, ed. *Risk and the War on Terror*. London: Routledge, 2007.

Anderson, G. et al. "Causes of the 2003 Major Grid Blackouts in North America and Europe, and Recommended Means to Improve System Dynamic Performance." *IEEE Transactions of Power Systems* 20, no. 4 (November 2005): 1922–28.

Anderson, M., and O. Suess. 2006. "The Allure of Catastrophe Bonds." *International Herald Tribune*, July 13, 2006.

Andreas, P., and R. Price. "From War Fighting to Crime Fighting: Transforming the American National Security State." *International Studies Association* 3, no. 3 (2001): 31–52.

Anjaria, J. "Urban Calamities: The View From Mumbai." *Space and Culture*, 9, part 1 (January, 2006): 80–92.

Ansary, T. "A War Won't End Terrorism." *San Francisco Chronicle*, October 19, 2002. http://www.commondreams.org/views02/1019-02.htm

———. "An Afghan-American Speaks," Salon.com, September 14, 2001. http://archive.salon.com/news/feature/2001/09/14/afghanistan/

Ansems de Vries, L. "(The War On) Terrorism: Destruction, Collapse, Mixture, Reenforcement, Construction." *Cultural Politics* 42 (2008): 183–98.

Arabian Business. "Dubai Opens First 'Luxury' Labour Camp." 2007.http://www.arabianbusiness.com/index.php?option=com_content&view=article&id=8473

Arey, J., and A. Wilder. "New Orleans Police Respond to Katrina." *Law and Order Magazine*, October 2005. http://www.hendonpub.com/resources/articlearchive/details.aspx?ID=33.

———. "NOPD SWAT Versus Katrina." *Law and Order Magazine* January 2006.http://www.hendonpub.com/resources/articlearchive/details.aspx?ID=199.

Associated Press. "Katrina's Devastation Reaches into the Psyches of Survivors." *Pittsburgh Tribune-Review*, September 7, 2005.

———. "Four Dead, Thousands Evacuated in Gaza Sewage Flood." March 27, 2007. http://www.iht.com/articles/ap/2007/03/27/africa/ME-GEN-Gaza-Sewage-Flood.php

———. "Malnutrition Common for Gaza Kids." *Jerusalem Post,* April 11, 2007. http://www.jpost.com/servlet/Satellite?pagename=JPost%2FJPArticle%2FShowFull&cid=1176152773887

———. "Blackout Inquiry Result Could Come by End of Week." *Roanoke Times* August 17, 2003, A1, 9.

Astore, W. "Attention Geeks and Hackers, Uncle Sam's Cyber Force Wants You!" *Tom Dispatch*, June 5, 2008. http://www.tomdispatch.com/post/174940/william_astore_militarizing_your_cyberspace

Auge, M. *Non-Places*. London: Verso, 1995.

Bak, P., and M. Paczuski. "Complexity, Contingency, and Criticality." *Proceedings of the National Academy of Sciences* 92 (1995): 6689–96.

Banerjee, N., and D. Firestone. "New Kind of Electricity Market Strains Old Wires Beyond Limits." *New York Times,* August 24, 2003, A1, 26.

Barakat, S. "City War Zones." *Urban Age* (Spring 1999): 11–19.

Barriot, P., and C. Bismuth. "Ambiguous Concepts and Porous Borders." In *Treating Victims of Weapons of Mass Destruction: Medical, Legal and Strategic Aspects,* edited by Patrick Barriot and Chantal Bismuth. London: Wiley, 2008.

Barry, A. "The Anti-Political Economy." *Economy and Society* 31, no. 2 (2002): 268–84.

Barry, D., and J. Longman. "A Police Department Racked by Doubt and Accusations." *The New York Times*, September 30, 2005.

Barry, J. M. *Rising Tide: The Great Mississippi Flood of 1927 and How It Changed America*. New York: Touchstone, 1997.

Bateson, G. *Steps to an Ecology of Mind*. New York: Ballantine, 1978.

Baudrillard, J. *For a Critique of the Political Economy of the Sign*. St. Louis, MO: Telos Press, 1981.

———. *The System of Objects*. London: Verso, 1996.

Baum, D. "Deluged: When Katrina Hit, Where Were The Police?" *The New Yorker*, January 9, 2006. http://www.newyorker.com/archive/2006/01/09/060109fa_fact

Beck, U. *The Risk Society: Towards a New Modernity*. London: Sage, 1992.

———. *Weltrisikogesellschaft*. Frankfurt: Suhrkamp, 2007.

Behnke, A. "The Re-Enchantment of War in Popular Culture." *Millennium: Journal of International Studies*, 34, no. 3 (2006): 937–49.

Behr, P. "Probers Say Blackout in August Was Avoidable." *Washington Post*, November 20, 2003, A1.

Bellflow, J. W. "The Indirect Approach." *Armed Forces Journal*. 2006. http://www.afji.com/2007/01/2371536

Benedikt, M. *Cyberspace: First Steps*. Cambridge, MA: MIT Press, 1992.

Bennett, C. "What Happens When You Book An Airline Ticket (Revisited): The Computer Assisted Passenger Profiling System and the Globalization of Personal Data." In *Global Surveillance and Policing: Borders, Security, Identity*, edited by Elia Zureik and Mark B. Salter, 113–38. Uffculme, Devon, UK: Willan, 2006.

Bennett, J. "The Agency of Assemblages and the North American Blackout." *Public Culture* 17, no. 3 (2005): 445–65.

Bigo, D. "To Reassure, and Protect, After September 11," Social Science Research Council, n.d. http://www.ssrc.org/sept11/essays/bigo.htm (accessed March 20, 2007).

Bijker, W., "Do Not Despair: There Is Life After Constructivism." *Science, Technology and Human Values* 18, no. 1 (1993): 113–38.

———. *Of Bicycles, Bakelites, and Bulbs: Toward a Theory of Sociotechnical Change*. Cambridge, MA: MIT Press, 1995.

———. T. Hughes, and T. Pinch, eds. *The Social Construction of Technological Systems: New Directions in the Sociology and History of Technology*. Cambridge, MA: MIT Press, 1987.

Blakeley, R. Bomb Now, Die Later. Bristol University, Department of Politics, 2003. http://www.geocities.com/ruth_blakeley/bombnowdielater.htm (accessed February 2004).

———. "Targeting Water Treatment Facilities." Campaign Against Sanctions in Iraq, Discussion List, January 24, 2003. http://www.casi.org.uk/discuss/2003/msg00256.html

Bolkcom, C., and J. Pike. (1993) *Attack Aircraft Proliferation: Issues for Concern*. Federation of American Scientists, 1993. http://www.fas.org/spp/aircraft (accessed February 25, 2004).

Bonacich, E. "Labor and the Global Logistics Revolution." In *Critical Globalization Studies*, 359–68. New York: Routledge, 2005.

Bourdieu, P. *Distinction: A Social Critique of the Judgement of Taste*. Cambridge, MA: Harvard University Press, 1984.

———. L. Wacquant. "The New Planetary Vulgate." In *Pierre Bourdieu Political Interventions: Social Science and Political Action*, selected and introduced by Franck Poupeau and Thierry Discepolo, 364–69. London: Verso, 2008.

Bowen, J. and C. Laroe. "Airline Networks and the International Diffusion of Severe Acute Respiratory Syndrome (SARS)." *The Geographical Journal* 172, no. 2 (2006): 130–44.

Bowker, G., and S. L. Star. *Sorting Things Out: Classification and Its Consequences*. Cambridge, MA: MIT Press, 2000.

Branford, S. "Food Crisis Leading to an Unsustainable Land Grab." *The Guardian*, November 22, 2008. http://www.guardian.co.uk/environment/2008/nov/22/food-biofuels

Brown, M. B Sovacool, and Richard Hirsh. "Assessing U.S. Energy Policy." *Daedalus* (Summer. 2006): 5–11.

Brown, W. "American Nightmare: Neoliberalism, Neoconservatism, and De-Democratization" *Political Theory* 34, no. 6 (2006): 690–714.

Busch, L. "Performing the Economy, Performing Science: From Neoclassical to Supply Chain Models in the Agrifood Sector." *Economy and Society* 3 (2007): 437–66.

Callon, M., ed. *The Laws of the Markets*. London: Blackwell, 1998.

———. J Law. "Guest Editorial: Introduction: Absence–Presence, Circulation, and Encountering in Complex Space." *Environment and Planning D: Society and Space* 22 (2004): 3–11.

Campbell, D. "The Biopolitics of Security: Oil, Empire, and the Sports Utility Vehicle." *American Quarterly* 57, no. 3 (2005): 943–72.

Canadian Consortium on Human Security. "Human Security > Cities." 2006. http://www.humansecurity-cities.org/page119.ht

Canadian Health Professionals. "Statement of Concern for the Public Health Situation in Gaza." Open Letter, July 31, 2006. http://electronicintifada.net/v2/article5357.shtml

Carroll, B. "Seeing Cyberspace: The Electrical Infrastructure as Architecture." In *Sarai Reader: Cities and Everyday Life*. Sarai: New Delhi, 2002. http://www.sarai.net/publications/readers/02-the-cities-of-everyday-life/04cyber_ electric.pdf (accessed January 2009).

Casre International. "Crisis in Gaza." 2006. http://www.care-international.org/index.php?option=com_content&task=view&id=97&Itemid=94

Castells, M. *The Informational City*. Oxford: Blackwell, 1989.

Cha, A. "New Orleans Police Keep Public Trust, Private Pain." *The Washington Post*, September 12, 2005.

Chatterjee, S. "Mumbai's 'Shanghai Dream' Hangs In Balance." Bombay's Demolition Drive, February 27, 2005. http://dupb.blogspot.com/2005_02_01_archive.html

Chelala, C. "Strangling Gaza." *Common Dreams*, December 15, 2007. http://www.commondreams.org/archive/2007/12/14/5844/

Church, W, "Information Warfare." *International Review of the Red Cross* 837 (2001): 205–16.

City of New York. "Waste Water Treatment: Preventing Discharges into Sewers: Guidelines for New York City Businesses." Department of Environmental Protection, City of New York, 2002. http://www.nyc.gov/html/dep/html/grease.html

Clarke, L. *Mission Improbable: Using Fantasy Documents to Tame Disaster*. Chicago: University of Chicago Press, 1999.

Coaffee, J, D Murakami Wood, and P Rogers. *The Everyday Resilience of the City: How Cities Respond to Terrorism and Disaster*. London: Palgrave Macmillan, 2008.

Collier, S., and A. Lakoff. "On Vital Systems Security." Berkeley, CA: Anthropology of the Contemporary Research Collaboratory, 2006. http://www.anthropos-lab.net (accessed March 29, 2007).

———. "How Infrastructure Became a Security Problem." In *The Changing Logics of Risk and Security*, edited by Myriam Dunn. New York: Routledge, forthcoming.

———. "The Vulnerability of Vital Systems: How 'Critical Infrastructure' Became a Security Problem." In *The Politics of Securing the Homeland: Critical Infrastructure, Risk and Securitisation*, edited by Myriam Dunn and Kristian Soby Kristensen. London and New York: Routledge, 2008.

Colten, C. *An Unnatural Metropolis: Wresting New Orleans from Nature*. Baton Rouge: Louisiana State University Press, 2005.

Columbia Accident Investigation Board. *Report,* Vol. 1. Washington, DC: U.S. Government Printing Office, 2003.

Congressional Research Service (CRS). "Border and Transportation Security: The Complexity of the Challenge." 2005. http://www.fas.org/sgp/crs/homesec/RL32839.pdf (accessed March 29, 2007).

Connolly, C. "Katrina's Emotional Damage Lingers." *The Washington Post*, December 7, 2005.

Cooke, J. "The DP World Controversy Could Help our Seaports." *Logistics Management*, 4, no. 1(2006).

Cooper, M. (2008) "Infrastructure and Event: Urbanism and the Accidents of Finance" Paper presented to CUNY Center for Place, Culture and Politics, New York, 2006.

Coutard, O, ed. *The Governance of Large Technical Systems*. London: Routledge, 1999.

Cowen, D. "National Soldiers and the War on Cities." *Theory & Event* 10 (2) (2007).

———.*Military Workfare: The Soldier and Social Citizenship in Canada*. Toronto: University of Toronto Press, 2008a.

———. "A Geography of Logistics: Reterritorializing Economy, Security and the Social." *Annals of the Association of American Geographers* (2008b).

———. Smith, S. "After Geopolitics? From the Geopolitical Social to Geoeconomics." *Antipode* (forthcoming).

Cresswell, T. *On the Move*. London: Routledge: London, 2006.

Curry, M. "The Profiler's Question and the Treacherous Traveller: Narratives of Belonging in Commercial Aviation." *Surveillance and Customs and Border Patrol (CBP)* 2006.

"Container Security Initiative, 2006–2011 Strategic Plan." http://www.customs.treas.gov/linkhandler/cgov/border_security/international_activities/csi/csi_strategic_plan.ctt/csi_strategic_plan.pdf (Accessed July 2, 2007).

Davis, M. *Buda's Wagon: A Brief History of the Car Bomb*. London: Verso, 2007.

———. *The Monster at Our Door*. New York and London: New Press.

Debrix, F. *Tabloid Terror: War, Culture and Geopolitics*. New York: Routledge, 2008.

Deer, P. "Introduction: The Ends of War and the Limits of War Culture." *Social Text* (2006): 25–36.

Delanda, M. *War in the Age of Intelligent Machines*. New York: Zone Books, 1991.

Deleuze, G., and F Guattari. "City/State." *Zone* 1–2 (1997): 195–99.

DeParle, J. "Fearful of Restive Foreign Labor, Dubai Eyes Reforms." *New York Times*, August 6, 2007.

Department of Homeland Security. "Fact Sheet: Securing US Ports." February 22, 2006. http://www.dhs.gov/xnews/releases/press_release_0865.shtm (accessed May 10, 2008).

———. *Strategy to Enhance International Supply Chain Security*. 2007. http://www.dhs.gov/xprevprot/publications/gc_1184857664313.shtm (accessed July 27, 2007).

Diamond, J. *Collapse: How Societies Choose to Fail or Survive*. London: Penguin, 2006.

Diken, B, and C Laustsen. "Camping as a Contemporary Strategy: From Refugee Camps to Gated Communities." *AMID Working Paper Series* 32 (2002): Aalborg University.

Dillon, M. "Virtual Security: A Life Science of (Dis)order." *Millennium* 32 (2003): 531–58.

Dodge, M., and R. Kitchin. "Flying Through Code/Space: The Real Virtuality of Air Travel." *Environment and Planning A* 36 (2004): 195–211.

Doganis, R. *The Airport Business*. New York: Routledge, 1992.

Douglas, M. *Purity and Danger: An Analysis of Concepts of Pollution and Taboo*. London: Routledge & Kegan Paul, 1966.

Department for Professional Employees (DPE). "Current Statistics on White Earthtimes. 2003: Liposuctioned Fat Could Be Bio-Diesel Fuel." http://www.earthtimes.org/articles/show/12174.html (accessed December 7, 2006).

Easterling, K. *Enduring Innocence*. Cambridge, MA: MIT Press, 2005.

Economist, The. "When Trade and Security Clash," April 4, 2002.

Edwards, P. "Infrastructure and Modernity: Force, Time, and Social Organization in the History of Sociotechnical Systems." In *Modernity and Technology*, edited by Thomas J. Misa, Philip Brey, and Andrew Feenberg, 185–225. Cambridge, MA: MIT Press, 2003.

Electronic Intifada. "Israel Invades Gaza: 'Operation Summer Rain'." http://electronicintifada.net/by-topic/442.shtml (accessed June 27, 2006).

Engel, M. "Land of the Fat." *The Guardian,* May 2, 2002. http://www.guardian.co.uk/world/2002/may/02/usa.medicalscience

Environmental Protection Agency. "Biodiesel: Fat to Fuel." http://epa.gov/region09/waste/biodiesel/index.html

———. "From Fat to Fuel; Santa Cruz Shares Results From First-In-Nation Community-Based Biodiesel Production." Press Release, 28 A, April28, 2008. http://yosemite.epa.gov/OPA/ADMPRESS.NSF/d0cf6618525a9efb85257359003fb69d/f012d2ab62a3bdde85257439006731ba!OpenDocument

Erikson, K. *Everything in Its Path: Destruction of Community in the Buffalo Creek Flood*. New York: Simon and Schuster, 1976.

Fackelmann, K. "Studies Tie Urban Sprawl to Health Risks, Road Danger: Driving Every ... Adds Up On Americans' Waistline, and Can Be Dangerous For Their Health." *USA Today*, August 29, 2003. http://www.usatoday.com/news/nation/2003-08-28-sprawl-usat x.htm

Fattah, H, and E Lipton. "Gaps in Security Stretch from Model Port in Dubai to U.S." *New York Times*, February 26, 2006.

Federation of American Scientists. n.d. "CBU-94 'Blackout Bomb' and BLU-114/B 'Soft-Bomb'." http://www.fas.org/man/dod-101/sys/dumb/blu-114.htm

Feldman, A. "Securocratic Wars of Public Safety." *Interventions: International Journal of Postcolonial Studies,* 6, no. 3 (2000): 330–50.

Felker, E. *Airpower, Chaos and Infrastructure: Lords of the Rings*. Maxwell Air Force Base, Alabama: U.S. Air War College Air University (Maxwell paper 14), 1998.

Fidler, D "The Globalization of Public Health: The First 100 Years of International Health Diplomacy." *Bulletin of the World Health Organization,* 79, no. 9 (2001): 842–49.

Filosa, G. "N.O. Police Chief Defends Force." *The Times-Picayune,* September 5, 2005.

Firestone, D, and R Pérez-Pe–a. "Failure Reveals Creaky System, Experts Believe." *New York Times,* August 15, 2003, A1, 18.

Flusty, S. *De-Coca-Colonization: Making the Globe from the Inside Out.* London and New York: Routledge, 2004.

Flynn, S. "The False Conundrum: Continental Integration versus Homeland Security." In *Rebordering North America,* edited by P. Andreas and T.J. Biersteker, 110–27. New York: Routledge, 2003.

Flynn, S. "The DP World Controversy and the Ongoing Vulnerability of U.S. Seaports." Testimony before the House Armed Services Committee. March 2, 2006. http://www.cfr.org/publication/9998/

Foster, M. "New Orleans Police Remain on Edge." *The Houston Chronicle,* October 11, 2005.

Foucault, M. *The Birth of the Clinic.* London: Tavistock, 1976.

———. *Language, Counter-Memory, Practice: Selected Essays and Interviews.* Edited by Donald F. Bouchard. Ithaca, NY: Cornell University Press, 1977.

———. *History of Sexuality,* Vol. 1. New York: Vintage, 1980.

———. *Technologies of the Self.* Amherst: University of Massachusetts Press, 1988.

———. *The Foucault Effect: Studies in Governmentality.* Edited by Graham Burchell, Colin Gordon, and Peter Miller. Chicago: University of Chicago Press, 1991.

———. *Security, Territory, Population: Lectures at the College de France 1977–1978.* Basingstoke, UK: Palgrave Macmillan, 2007.

Gaber, W, and R Gottschalk. "Entry- and Exit-Screening Procedures at Frankfurt International Airport." Powerpoint presentation, May 20, 2007.

Galloway, G. "WHO Concedes Advisory Damaged Toronto." *The Globe and Mail,* May 1, 2003, 1–2.

Gandy, M. *Concrete and Clay: Reworking Nature in New York City.* Cambridge, MA: MIT Press, 2002.

———. "Cyborg Urbanization: Complexity and Monstrosity in the Contemporary City." International *Journal of Urban and Regional Research* 29, no. 1 (March 2005): 26–49.

———. "The Paris Sewers and the Rationalization of Urban Space." *Transactions of the Institute of British Geographers* 24 (1999): 23–44

Garvey, J. *The Ethics of Climate Change: Right and Wrong in a Warming World.* London: Continuum, 2008.

Giacaman, R, and A Husseini. "Life and Health During the Israeli Invasion of the West Bank: The Town of Jenin." *Indymedia Israel* May 22, 2002.

Giddens, A. *The Nation-State and Violence.* Berkeley and Los Angeles: University of California Press, 1985.

Giedion, S. 1948. *Mechanization Takes Command: A Contribution to Anonymous History.* New York: Norton.

Gilbert, E. "Eye to Eye: Biometrics, the Observer, the Observed and the Body Politic." In *Observant States: Geopolitics and Visual Culture,* edited by F. MacDonald, K. Dodds, and R. Hughes. London: Tauris, forthcoming.

Gilman, N. *Mandarins of the Future: Modernization Theory in Cold War America.* Baltimore: Johns Hopkins, 2003.

Glanz, J. "Its Coils Tighten, and the Grid Bites Back." *New York Times,* August 17, 2003, A1, 4.

Glasser, S and M Grunwald. "The Steady Buildup to a City's Chaos." *The Washington Post,* September 11, 2005.

Goel, V. "What Do We Do with the SARS Reports?" *Healthcare Quarterly* 7, no. 3 (2004): 28–41.

Goffman, E. *The Presentation of the Self in Everyday Life.* Garden City, NY: Doubleday, 1959.

Goss, T. " 'Who's in Charge?' New Challenges n Homeland Defense and Homeland Security." *Homeland Security Affairs* 2 (2006): 1–12. http://www.hsaj.org/pages/volume2/issue1/pdfs/2.1.2.pdf

Gostin, L., R Bayer, and A Fairchild. "Ethical and Legal Challenges Posed by Severe Acute Respiratory Syndrome: Implications for the Control of Severe Infectious Disease Threats." *Journal of the American Medical Association* 290, no.24 (2003): 3229–37.

Gottdiener, M. *Life in the Air*. Oxford: Rowman and Littlefield, 2001.

Gottschalk, R. "Katastrophenschutz an internationalen Flughäfen" Contribution to Public Health Forum, Public Health bei Katastrophenfällen, 2006.

Gottschalk, R, and W Preiser. "Bioterrorism: Is It a Real Threat?" *Medical and Microbiological Immunology* 194 (2005): 109–14.

Graham, S. "Constructing Premium Networked Spaces: Reflections on Infrastructure Networks and Contemporary Urban Development." *International Journal of Urban and Regional Research* 24, no. 1 (2000a): 183–200.

———. "Lessons in Urbicide." *New Left Review* 19 (January–February 2003): 63–73.

———. "Beyond the 'Dazzling Light': From Dreams of Transcendence to the 'Remediation' of Urban Life." *New Media and Society* 6, no. 1 (2004): 16–26.

———. ed. *Cities, War and Terrorism*. Oxford: Blackwell, 2004.

———. "Software-Sorted Geographies." *Progress in Human Geography* 29, no. 5 (2005): 562–80.

———. "Switching Cities Off: Urban Infrastructure and U.S. Air Power." *City* 9, no. 2 (July 2005): 170–94.

———. "'Homeland' Insecurities? Katrina and the Politics of Security in Metropolitan America." *Space and Culture* 9, no. 1 (2006): 63–67.

———. *Cities Under Siege: The New Military Urbanism*. London: Verso, 2009.

———. Mike Crang, "Sentient Cities: Ambient Intelligence and the Politics of Urban Space." *Information, Communications and Society* 10, no. 6 (2008): 789–817.

———. Simon Marvin. "Telematics and the Convergence of Urban Infrastructure: Implications for Contemporary Cities." *Town Planning Review* 65, no. 3 (1994): 227–42.

———. *Splintering Urbanism: Network Infrastructures, Technological Mobilities and the Urban Condition*. London: Routledge, 2001.

———. Nigel Thrift. "Out of Order: Understanding Maintenance and Repair." *Theory, Culture and Society* 24, no. 3 (2007): 1–25.

Greenfeld, L. *The Spirit of Capitalism: Nationalism and Economic Growth*. Cambridge, MA: Harvard University Press, 2001.

Gregory, D "'In Another Time-Zone, the Bombs Fall Unsafely...': Targets, Civilians and Late Modern War." *Arab World Geographer* 9, no. 2 (2006): 88–111.

———. "Defiled Cities." *Singapore Journal of Tropical Geography* 24, no. 3 (2003): 307–26.

———. *The Colonial Present*. Oxford: Blackwell, 2004.

Groll-Yaari, Y, and H Assa. *Diffused Warfare: The Concept of Virtual Mass*. Haifa, February 2007 23.

Hables Gray, C. *The Cyborg Handbook*. New York: Routledge, 1995.

Haimes, Y. "Total Risk Management." *Risk Analysis*, 11, no. 2 (1991): 169–71.

Haraway, D. "A Manifesto For Cyborgs: Science, Technology, and Socialist-Feminism in the Late Twentieth Century." In *Simians, Cyborgs and Women: The reinvention of Nature*, ed. D. Haraway, 149–81. New York: Routledge, 1991.

Harvey, D. *The Urbanization of Capital*. Blackwell: Oxford, 1985.

———. "Globalization and the Body." In *Possible Urban Worlds*, edited by INURA. Basel: Birkhäuser, 1998.

———. "The City as Body Politic." In *Wounded Cities: Destruction and Reconstruction in a Globalized World*, edited by Jane Schneider and Ida Susser, 25–46. Oxford: Berg, 2003.

———. *A Brief History of Neoliberalism*. Oxford: Oxford University Press, 2005.

Haveman, J. & H. Shatz, eds. *Protecting the Nation's Seaports: Security and Cost*. San Francisco: Public Policy Institute of California, 2006.

Henke, C. "The Mechanics of Workplace Order: Toward a Sociology of Repair." *Berkeley Journal of Sociology* 43 (2000): 55–81.

Herman, R, and J Ausubel. *Cities and Their Vital Systems—Infrastructure Past, Present and Future.* Washington, DC: National Academy Press, 1988.

Heymann, D, M Kindhauser, and Rodier. "Coordinating the Global Response." In *SARS: How a Global Epidemic Was Stopped,* edited by WHO, 243–54. Geneva: World Health Organization, 2005.

Heynen, Nick, M Kaika, and E Sywyngedouw, eds. In *The Nature of Cities: Urban Political Ecology and the Politics of Urban Metabolism.* London: Routledge, 2006.

Hinchcliffe, S. "Technology, Power and Space—The Means and Ends of Geographies of Technology." *Environment and Planning D: Society and Space* 14, (1996): 659–82.

Hinkson, J. "After the London Bombings." *Arena Journal* 24 (2005): 139–59.

Hinman, E. "The Politics of Coercion." *Toward a Theory of Coercive Airpower for Post-Cold War Conflict* (CADRE Paper No. 14). Maxwell Air Force Base, Alabama: Air University Press, 36112-6615, August 2002.

Hirsh, M, and D Klaidman. "What Went Wrong." *Newsweek,* 142, no. 8, (August 25, 2003): 33–39.

Hirsh, R. *Power Loss: The Origins of Deregulation and Restructuring in the American Electric Utility System.* Cambridge, MA: MIT Press, 1999.

Hobsbawm, E. "The Future of War and Peace." *CounterPunch,* February 27, 2002. http://www.counterpunch.org/hobsbawm1.html

Hommels, A. *Unbuilding Cities: Obduracy in Urban Socio-Technical Change.* Cambridge, MA: MIT Press, 2005.

Hustmyre, C. "NOPD Versus Hurricane Katrina." *Tactical Response Magazine,* July 2006. http://www.hendonpub.com/resources/articlearchive/details.aspx?ID=745

Iklé, F. *Annihilation from Within: The Ultimate Threat to Nations.* New York: Columbia University Press, 2006.

Independent Levee Investigation Team (ILIT). *Investigation of the Performance of the New Orleans Flood Protection Systems in Hurricane Katrina on August 29, 2005.* 2006. http://www.ce.berkeley.edu/~new_orleans/ (accessed May 7, 2008).

Institute of Medicine (IOM). *To Err Is Human: Building a Safer Health System.* Washington, DC: National Academy Press, 2000.

Insurance Journal. "S&P Rates Mexican Quake Cat Bond Managed by Swiss Re 'BB+'." (May 22, 2006).

Interagency Performance Evaluation Task Force (IPET). *Performance Evaluation of the New Orleans and Southeast Louisiana Hurricane Protection System.* March 26, 2007. https://ipet.wes.army.mil/ (accessed May 2, 2008).

Isin, E. *Being Political: Genealogies of Citizenship.* Minneapolis: University of Minnesota Press, 2003.

Jackson, R "The Impact of the Built Environment on Health: An Emerging Field." *American Journal of Public Health,* 93 (2003): 1382–84.

Jameson, F. *Postmodernism, or the Cultural Logic of Late Capitalism.* Durham, NC: Duke University Press, 1991.

Johnson, B. "NATO Says Cyberwarfare Poses as Great a Threat as a Missile Attack." *The Guardian,* March 6, 2008. http://www.guardian.co.uk/technology/2008/mar/06/hitechcrime.uksecurity

Joy, W. "Why the Future Doesn't Need Us." *Wired* (April 2000): 238–60.

Kaika, M. "Interrogating the Geographies of the Familiar: Domesticating Nature and Constructing the Autonomy of the Modern Home." *International Journal of Urban and Regional Research,* 28, no. 2 (2004): 265–86.

———. *City of Flows,* London: Routledge, 2004.

———. Swyngedouw, E. "Fetishising the Modern City: The Phantasmagoria of Urban Technological Networks." *International Journal of Urban and Regional Research* 24, no. 1 (2000): 122–48.

Kaplan, S., and B. J. Garrick. "On the Quantitative Assessment of Risk." *Risk Analysis* 1, no.1 (1981): 11–27.

Keaney, T, and E Cohen. *Gulf War Air Power Surveys* (GWAPS), Vol. 2, Part 2. Washington, DC: Johns

Hopkins University and the U.S. Air Force, 1993. http://www.au.af.mil/au/awcgate/awc-hist. htm+gulf (accessed February 15. 2004).

Keil, R. *Los Angeles: Globalization, Urbanization, and Social Struggles.* Chichester, UK: Wiley, 1998.

———. S. Ali. "Governing the Sick City: Urban Governance in the Age of Emerging Infectious Disease." *Antipode* 40, no. 1 (2007): 846–71.

———. "Multiculturalism, Racism and Infectious Disease in the Global City." *Topia: Canadian Journal of Cultural Studies* 16 (2006): 23–49.

———. "The Avian Flu: The Lessons Learned from the 2003 SARS Outbreak in Toronto." *Area* 38, no. 1 (2006): 107–9.

Killingsworth, R, j Earp, and R Moore. "Introduction: Supporting Health Through Design: Challenges and Opportunities." *American Journal of Health Promotion* (September/October 2003): 1–3.

Kimber, I. "What Happened to the Gaza Strip?" *IMEMC News*, October 13, 2006. http://www.imemc. org/content/view/22059/117/

Klein, N. *The Shock Doctrine: The Rise of Disaster Capitalism.* London: Penguin, 2007.

Kletz, T. *Learning From Accidents.* Oxford, UK: Gulf Professional, 2001.

Klinenberg, E. *Heat Wave: A Social Autopsy of Disaster in Chicago.* Chicago: University of Chicago Press, 2006.

Koning-AbuZayd, K. "This Brutal Siege of Gaza Can Only Breed Violence in Gaza City. *The Guardian*, January 23, 2008. http://www.guardian.co.uk/commentisfree/2008/jan/23/israelandthepalestinians.world

Konvitz, J. "Why Cities Don't Die." *American Heritage of Invention and Technology* (Winter 1990): 62.

Kostof, S. *The City Assembled.* London: Thames and Hudson, 1992.

Kraska, P. *Militarizing the American Criminal Justice System: The Changing Roles of the Armed Forces and the Police.* Boston: Northeastern University Press, 2001.

La Porte, T. "A Strawman Speaks Up: Comments on *The Limits of Safety.*" *Journal of Contingencies and Crisis Management* (1994): 207–11.

———. P. Consolini. "Working in Practice But Not in Theory: Theoretical Challenges of High Reliability Organizations." *Journal of Public Administration Research and Theory* (Winter 1991): 19–47.

———. G. Rochlin. "A Rejoinder to Perrow." *Journal of Contingencies and Crisis Management* (1994): 221–27.

Lake, A., and T. Townshend. "Obesogenic Environments: Exploring the Built and Food Environments." *Perspectives in Public Health* 126 (2006): 262–67.

Lakoff, A. "Preparing for the Next Emergency." *Public Culture* 19, no. 2 (2007): 247–71.

Latour, B. "Where are the Missing Masses? The Sociology of a Few Mundane Artifacts." In *Shaping Technology/Building Society*, edited by W. E. Bijker and J. Law, 225–58. Cambridge, MA: MIT Press, 1992.

———. *We Have Never Been Modern.* London: Harvester and Wheatsheaf, 1993.

———. *Aramis, Or the Love of Technology.* Cambridge, MA: Harvard University Press, 1996.

———. *Pandora's Hope: Essays on the Reality of Science Studies.* Cambridge, MA: Harvard University Press, 1999.

Lau, S., K. Woo, Y. Li, H. Huang, B. Tsoi, S. Wong, K. Leung, Y. Chan, and K. Yuen. "Severe Acute Respiratory Syndrome Coronavirus-Like Virus in Chinese Horseshoe Bats." *Proceedings of the National Academy of Sciences of the United States* 102, no. 39 (2005): 14040–45.

Law, J, and J Hassard, eds. *Actor Network Theory and After.* London: Wiley, 1999.

Leslie, J. "Powerless." *Wired* (April 1999): 119–83.

Li, D. "The Gaza Strip as Laboratory: Notes in the Wake of Disengagement." *Journal of Palestine Studies* 35, no. 2 (2006): 38–55.

Liang, Q., and W. Xiangsui. *Unrestricted Warfare.* Panama: Pan American, 2002.

Little, R. "Controlling Cascading Failure: Understanding the Vulnerabilities of Interconnected Infrastructures." *Journal of Urban Technology* (2002): 109–23.

———. "A Socio-Technical Systems Approach to Understanding and Enhancing the Reliability of Interdependent Infrastructure Systems." *International Journal of Emergency Management* 2, no. 1–2 (2004): 98–110.

Loewenstein, J. "Notes from the Field: Return to the Ruin that is Gaza." *Journal of Palestine Studies* 36, no. 3 (Spring 2007): 23–35.

Luke, Timothy. *Screens of Power: Ideology, Domination, and Resistance in Informational Society.* Urbana: University of Illinois Press, 1989.

———. "Identity, Meaning and Globalization: Space-Time Compression and the Political Economy of Everyday Life." In *Detraditionalization: Critical Reflections on Authority and Identity*, edited by S Lash, P Heelas, and P Morris, 109–33. Oxford: Blackwell, 1996.

———. *Capitalism, Democracy, and Ecology: Departing from Marx.* Urbana: University of Illinois Press, 1999.

———. "Real Interdependence: Discursivity and Concursivity in Global Politics." *Language, Agency and Politics in a Constructed World*, edited by F Debrix. Armonk, NY: M. E. Sharpe, 2003.

———. "The Coexistence of Cyborgs, Humachines, and Environments in Postmodernity: Getting Over the End of Nature." *The Cybercities Reader*, edited by S. Graham. Oxford: Blackwell, 2003.

———. "Everyday Technics as Extraordinary Threats: Urban Technostructures and Nonplaces in Terrorist Actions." In *Cities, War and Terrorism*, edited by Stephen Graham, 120–36. Oxford: Blackwell, 2004.

MacKenzie, D. *Inventing Accuracy: A Historical Sociology of Nuclear Missile Guidance.* Cambridge, MA: MIT Press, 1990.

MADRE. *War on Civilians: A MADRE Guide to the Middle East Crisis,* July 19, 2006. http://www.madre.org/articles/me/waroncivilians.html

Major, C. "Affect Work and Infected Bodies: Biosecurity in an Age of Emerging Infectious Disease." *Environment and Planning A* 40, no.7 (2008): 1633–46.

Marburger, J. Testimony before the House Committee on Science. June 14, 2002.

Marquand, R., and B. Arnoldy. "China Emerges as Leader In Cyberwarfare." *Christian Science Monitor,* September 14, 2007. http://www.csmonitor.com/2007/0914/p01s01-woap.html

Marx, K. *Capital: A Critique of Political Economy,* Vol. 2, *The Process of Circulation of Capital.* Chicago: Charles H. Kerr, 1909.

———. *Grundrisse,* translated by Martin Nicholaus. New York: Vintage, 1973.

———. F. Engels. "The Communist Manifesto." In *The Marx-Engels Reader*, edited by Robert C. Tucker. New York: Norton, 1978.

Mau, B. *Massive Change.* London: Phaidon, 2003.

Mayer, J. "Geography, Ecology and Emerging Infectious Diseases." *Social Science and Medicine* 50 (2000): 937–52.

Mayntz, R. "Technological Progress, Societal Change and the Development of Large Technical Systems." Mimeo, 1995.

McCarthy, R. "Hamas Offers $52m Handouts to Help Hardest-Hit Gazans." *The Guardian,* January 26, 2009. http://www.guardian.co.uk/world/2009/jan/26/hamas-payout-gaza-infrastructure

McFarlane, C., and J. Rutherford. "Political Infrastructures: Governing and Experiencing the Fabric of the City." *International Journal of Urban and Regional Research* 32, no. 2 (June 2008): 363–74.

McGinn, D., and K. Naughton. "The Price of Darkness." *Newsweek*, 142, no. 8 (August 25, 2003): 42–43.

McQuaid, J. "Alarm Sounded Too Late as N.O. Swamped—Slow Response Left City in Lurch." *The Times-Picayune,* September 8, 2005.

Methvin, E. "Undercover Against the Mob." *Reader's Digest,* May 2000.

Metz, S. "The Next Twist of the RMA." *Parameters* (Autumn 2000): 40–53. http://www.carlisle.army.mil/usawc/Parameters/00autumn/metz.htm

Mileti, D. S. *Disaster by Design: A Reassessment of Natural Hazards in the United States.* Washington, DC: Joseph Henry Press, 1999.

Milne, D. "'Our Equivalent of Guerrilla Warfare': Walt Rostow and the Bombing of North Vietnam, 1961–1968." *The Journal of Military History* 71 (2007): 169–203.

Milne, S. "To Blame the Victims for This Killing Spree Defies Both Morality and Sense." *The Guardian,* March 5. 2008. http://www.guardian.co.uk/commentisfree/2008/mar/05/israelandthepalestinians.usa

Mitchell, W., and A. Townsend. "Cyborg Agonistes: Disaster and Reconstruction in the Digital Era." In *The Resilient City: How Modern Cities Recover from Disaster*, edited by Lawrence Vale and Thomas Campanella, 313–24. Oxford, UK: Oxford University Press, 2005.

Mumford, L. *Technics and Civilization.* New York: Harcourt Brace Jovanovich, 1963.

Murakami Wood, D., and J. Coaffee. "Security is Coming Home: Rethinking Scale and Constructing Resilience in the Global Urban Response to Terrorist Risk." *International Relations* 20, no. 4 (2006): 503–17.

National Advisory Committee on SARS and Public Health (NACSPH). *Learning from SARS: Renewal of Public Health in Canada.* Ottawa: Health Canada, 2003.

National Research Council (NRC). *Understanding Risk: Informing Decisions in a Democratic Society.* Washington, DC: National Academy Press, 1996.

Newman, B. "The Sewer–Fat Crisis Stirs a National Stink: Sleuths Probe Flushing." *The Wall Street Journal,* June 4, 2001. http://www.viridiandesign.org/notes/251-300/00257_sewer_fat_crisis.html

New York Times. "Securing the Flow of Goods." February 26, 2006. http://www.nytimes.com/imagepages/2006/02/26/national/nationalspe cial3/20060226_PORT_GRAPHIC.html

North American Electric Reliability Council (NAERC). *August 14, 2003 Blackout: NAERC Actions to Prevent and Mitigate the Impacts of Future Cascading Blackouts.* Princeton, NJ: NAERC, 2004.

Nye, D. *Electrifying America: Social Meanings of a New Technology.* Cambridge, MA: MIT Press, 1990.

———. *The Technological Sublime.* Cambridge, MA: MIT Press, 1996.

Ong, A., and S. Collier, eds. *Global Assemblages: Technology, Politics, and Ethics as Anthropological Problems.* Oxford: Blackwell, 2005.

Onley, D. "U.S. Aims to Make War On Iraq's Networks." Missouri Freedom of Information Center, 2003. http://foi.missouri.edu/terrorbkgd/usaimsmake.html (accessed February24, 2004).

Organization for Economic Development and Cooperation (OECD). *Security in Maritime Transport: Risk Factors and Economic Impact: Maritime Transport Committee Report.* 2003.

O'Rourke, T., A. Lembo, and L. Nozick. "Lessons Learned from the World Trade Center Disaster about Critical Utility Systems." In *Impacts of and Human Response to the September 11, 2001 Disasters: What Research Tells Us*, edited by M. F. Myers, 269–90. Boulder, CO: Natural Hazards Research and Applications Information Center, University of Colorado, 2003.

Pagano, V. "Letter to Nitin Patel, RE: Zircon Industries, Inc." NYC Department of Environmental, Protection Bureau of Water and Sewer Operations, June 30, 1999. http://www.greenchem.com/cityofnewyork.html

Page, B. "Paying for Water and the Geography of Commodities." *Transactions of the Institute of British Geographers* 30, no. 3 (September, 2005): 293–306.

Page, M. *The City's End.* New Haven, CT: Yale University Press, 2008.

Palestinian Medical Relief Society. Public Health Disaster in Gaza Strip: Urgent Appeal for Support to Avert Public Health Disaster in the Gaza Strip, June 27, 2007. http://www.pmrs.ps/last/etemplate.php?id=18

Park, R., E. Burgess, and R. McKenzie. *The City.* Chicago: Chicago University Press, 1967.

Parlette, V, and Cowen, D. "Dead Malls: Old Suburbs and New Logistic Spaces." (Unpublished paper, 2008, available from the authors. deb.cowen.cowen@utoronto.ca)

Pawley, M. *Terminal Architecture.* London: Reaktion Books, 1997.

Pérez-Pe–a, R. "Utility Could Have Halted '03 Blackout, Panel Says." *New York Times*, April 6, 2004, A16.

————. M Wald. "Basic Failures by Ohio Utility Set Off Blackout, Report Finds." *New York Times,* November 20, 2003, A1, 28.

Perlstein, M, and T Lee. "The Good and the Bad." *The Times-Picayune,* December 18, 2005. http://www.nola.com/crime/t-p/index.ssf?/crime/stories/good_bad.html

Perrow, C. "The Limits of Safety: The Enhancement of a Theory of Accidents." *Journal of Contingencies and Crisis Management* (1994): 212–20.

————. *Normal Accidents: Living with High-Risk Technologies.* Princeton, NJ: Princeton University Press, 1999a.

————. "The Vulnerability of Complexity." Paper presented at the Planning Meeting on the Role of the National Academies in Reducing the Vulnerabilities of Critical Infrastructures held at the National Academy of Sciences, April 28–29, Washington, DC, 1996.

Perry, D. "Introduction." In D. Perry, *Building the Public City: The Politics, Governance and Finance of Public Infrastructure,* 1–20. London: Sage, 1995.

Petroski, H. *To Engineer Is Human: The Role of Failure in Successful Design.* New York: Vintage Books, 1992.

————. 1994. *Design Paradigms: Case Histories of Error and Judgment in Engineering.* Cambridge, UK: Cambridge University Press.

Phimister, J., Oktem, U., Kleindorfer, P., and Kunreuther, H. "Near-Miss System Analysis: Phase I." Working Paper of the Near-Miss Project. Wharton School, Center for Risk Management and Decision Processes, 2000. http://www.opim.wharton.upenn.edu/risk/proj/nearmiss.html

Picon, A. *La Ville Territoire de Cyborgs.* Paris: L'Imprimeur, 1998.

Pike, D. *Metropolis on the Styx: The Underworlds of Modern Urban Culture, 1800–2001.* New York: Columbia University Press, 2007.

Pineiro, R. *Cyberterror,* New York: Forge Press, 2003.

Pool, R. *Beyond Engineering: How Society Shapes Technology.* Oxford, UK: Oxford University Press, 1997.

Pursell, C. *The Machine in America: A Social History of Technology.* Baltimore: Johns Hopkins University Press, 1995.

Rabinbach, A. *The Human Motor: Energy, Fatigue, and the Origins of Modernity.* New York: Basic, 1990.

Rammert, W. "New Rules of Sociological Method: Rethinking Technology Studies." *British Journal of Sociology* 48 (1997): 171–91.

Rao, V. "How to Read a Bomb: Scenes from Bombay's Black Friday." *Public Culture,* 19, no.3 (2007): 567–92.

Rattray, G. *Strategic Warfare in Cyberspace.* Cambridge, MA: MIT Press, 2001.

Redhead, S. "The Art of the Accident: Paul Virilio and Accelerated Modernity." *Fast Capitalism,* 2, no. 1 (2006). http://www.uta.edu/huma/agger/fastcapitalism/2_1/redhead.htm

Reid, J. "Open Source Destruction, Weak Discipline, War Infrastructure." In *Under Fire,* edited by Jordan Crandall. http://underfire.eyebeam.org/?q=node/462 (accessed February 1, 2009).

Revill, J. "A Deadly Slice of American Pie." *The Observer,* September 21, 2003, 17. http://www.guardian.co.uk/society/2003/sep/21/health.lifeandhealth

Reynolds, D. "Your City's Fat—Now What? Being America's Fattest City Isn't Good For One's Image." *ABC News,* January 4, 2002. http://abcnews.go.com/sections/wnt/worldnewsTonight/ fatcity020104.htm

Rhoades, D. L. *Evolution of International Aviation.* Burlington VT: Ashgate, 2003.

Rinaldi, S., J. P. Peerenboom, and T. K. Kelly. "Identifying, Understanding, and Analyzing Critical Infrastructure Interdependencies." *IEEE Control Systems Magazine* (December 2001): 11–25.

Rizer, K. "Bombing Dual-Use Targets: Legal, Ethical, and Doctrinal Perspectives." Air and Space Power Chronicles, January 5, 2001. http://www.airpower.maxwell.af.mil/airchronicles/cc/Rizer.html

Robb, J. "Cascading System Failure." May 24, 2004 http://globalguerrillas.typepad.com/ (accessed January 21, 2009).

———. *Brave New War: The Next Stage of Terrorism and the End of Globalization.* New York: Wiley, 2007.

Rochlin, G. "Networks and the Subversion of Choice: An Institutionalist Manifesto." *Journal of Urban Technology* 8, no. 3 (2001): 65–96.

———. *Trapped in the Net: The Unanticipated Consequences* of Computerization. Princeton, NJ: Princeton University Press, 1997.

Rose, N. *Governing the Soul: The Shaping of the Private Self*, 2nd ed. London: Free Association Books, 1999.

Rosenberg, B. "Cyber Warriors: USAF Cyber Command Grapples with New Frontier Challenges." *C4ISR Journal* (August 1, 2007).

Rostow, W. *The Stages of Economic Growth: A Non-Communist Manifesto*, Cambridge, UK: Cambridge University Press, 1960.

Rowat, C. 'Iraq Potential Consequences of War." *Campaign against Sanctions in Iraq Discussion List* November 8 (2003). http://www.casi.org.uk/discuss/2002/msg02025.html (accessed February 12, 2004).

Rumsfeld, D. News Transcript, U.S. Department of Defense (March 22, 2004). http://www.defenselink.mil/transcripts/transcript.aspx?transcriptid=2361

Rygiel, K. (2009) "The Securitized Citizen." In *Recasting the Social in Citizenship*, edited by Engin Isin, Toronto: University of Toronto Press.

Sagan, S. *The Limits of Safety.* Princeton, NJ: Princeton University Press, 1999.

Salehi, R., and S. H. Ali. "The Social and Political Context of Disease Outbreaks: The Case of SARS in Toronto." *Canadian Public Policy* 23, no. 4 (2006): 573–86.

Salter, M. "The Global Visa Regime and the Political Technologies of the International Self: Borders, Bodies, Biopolitics." *Alternatives* 31(2006): 167–189.

Sarasin, P. *Anthrax: Bioterror as Fact and Fantasy.* Cambridge, MA: Harvard University Press, 2006.

Sassen, S. *Globalization and Its Discontents: Essays on the New Mobility of People and Money.* New York: New Press, 1998.

Schneider, J., and I. Susser, eds. *Wounded Cities: Destruction and Reconstruction in a Globalized World.* London: Berg.

Sennett, R. "The Age of Anxiety." *Guardian*, October 23, 2004. http://books.guardian.co.uk/review/story/0,,1332840,00.html

Serres, M. *Angels: A Modern Myth.* New York: Flammarion, 1995.

Shaoul, J. "The Plight of the UAE's Migrant Workers: The Flipside of a Booming Economy." (November 9, 2007). http://www.wsws.org/articles/2007/nov2007/duba-n09.shtml

Shell, E. *Fat Wars: The Inside Story of the Obesity Industry.* London: Atlantic, 2002.

Sims, B. "'The Day after the Hurricane': Infrastructure, Order, and the New Orleans Police Department's Response to Hurricane Katrina." *Social Studies of Science* 37, no. 1 (February 2007): 111–18.

Sims, B. "Things Fall Apart: Disaster, Infrastructure, and Risk." *Social Studies of Science* 37, no. 1 (February 2007): 93–95.

Skoric, I. "On Not Killing Civilians." (May 6, 1999). http://www.amsterdam.nettime.org (accessed February16, 2004).

Smith, A. *The Wealth of Nations.* London: Penguin, 1987.

Smith, C. "U.S. Wrestles with New Weapons." *NewsMax.Com* (March 13, 2003). http://www.newsmax.com/archives/articles/2003/3/134712.shtml

Smith, N. *The Endgame of Globalization.* New York: Routledge, 2005.

Smith, R. "World City Topologies." *Progress in Human Geography* 27(2003): 561–82.

Smith, T. "The New Law of War: Legitimizing Hi-Tech and Infrastructural Violence." *International Studies Quarterly* 46 (2002): 355–74.

Smykay, E., and B. LaLonds. *Physical Distribution: The New and Profitable Science of Business Logistics.* Chicago, London: Dartnell Press, 1967.

Society 1, no.4 (2004): 475–99.

Solis, J. *New York Underground.* New York: Routledge, 2007.

Sorkin, M. "Urban Warfare: A Tour of the Battlefield." In *Cities, War and Terrorism: Towards an Urban Geopolitics,* edited by Stephen Graham, 251–62. Oxford: Blackwell, 2004.

"Southbound 101 Fully Reopens After Grease Spill." *Sfgate* (January 7 2004). http://www.sfgate.com

Southerland, R. "Sewer Fitness: Cutting the Fat." *American City and County* (October 1, 2002). http://americancityandcounty.com/mag/government_sewer_fitness_cutting/index.html

Speak, S. and S. Graham, "Service Not Included: Marginalised Neighbourhoods, Private Service Disinvestment, and Compound Social Exclusion." *Environment and Planning A* (2000): 1985–2001.

Sponster, C. "Beyond the Ruins: The Geopolitics of Urban Decay and Cybernetic Play." *Science Fiction Studies* 20, no. 2 (1992): 2251–65.

Squires, G., and C. Hartman, eds. *There Is No Such Thing as a Natural Disaster: Race, Class and Hurricane Katrina.* New York: Routledge, 2006.

Srivastava, B. *Aviation Terrorism.* New Delhi: Manas, 2002.

Star, S. L. "The Ethnography Of Infrastructure." *American Behavioral Scientist,* 43, no. 3 (1999): 377–91.

Star, S. L., and K. Ruhleder. "Steps Toward an Ecology of Infrastructure: Design and Access for Large Information Spaces." *Information Systems Research* 7, no. 1 (1996): 111–134.

Steinberg, P., and R. Shields, eds. *What is a City? Rethinking the Urban After Hurricane Katrina.* Athens and London: University of Georgia Press, 2008.

Stone, B. "How to Fix the Grid." *Newsweek,* 142, no. 8 (August 25, 2003): 38.

Stone, P. "Space Command Plans For Computer Network Attack Mission." U.S. Department of Defense: Defense Link (January 14, 2003). http://www.defenselink.mil (accessed February 22, 2004)

Strand, G. Keyword: Evil," *Harper's Magazine,* March 2008, 64–65.

Sudjic, D. *The Edifice Complex: How the Rich and Powerful Shape the World.* London: Penguin, 2006.

Sugden, J. "Doctor Used 'Human Fat To Power Car'." *Times Online* December 24, 2008. http://www.timesonline.co.uk/tol/news/environment/article5393763.ece

Sui, D. "Musing on the Fat City: Are Obesity and Urban Forms Linked?" *Urban Geography* 24 (2003): 75–84.

Summerton, J. "Social Shaping in Large Technical Systems." *Flux,* 17 (July–September 1994): 54–56.

———. ed. *Changing Large Technical Systems.* Boulder, CO: Westview Press, 1994.

Sweet, K. *Aviation and Airport Security: Terrorism and Safety Concerns.* Upper Saddle River, NJ: Pearson Prentice Hall, 2004.

Swyngedouw, E. "Communication, Mobility and the Struggle for Power Over Space." In *Transport and Communications in the New Europe,* edited by G. Giannopoulos and A. Gillespie, 305–25. London : Belhaven, 1993.

Taleb, N. *The Black Swan: The Impact of the Highly Improbable.* New York: Random House, 2007.

Talukdar, S., J. Apt, M. Ilic, L. Lave, and M. Morgan. "Cascading Failures: Survival versus Prevention." *The Electricity Journal* (November 2003): 25–31.

Tierney, T. *The Value of Convenience: A Genealogy of Technical Culture.* Albany: SUNY Press, 1993.

Todd, A., and A. Wood. "'Flex Your Power': Energy Crises and the Shifting Rhetoric of the Grid." *Atlantic Journal of Communication* 14, no. 4 (2006): 211–28.

Torrance, M. "Forging Glocal Governance? Urban Infrastructures as Networked Financial Products." *International Journal of Urban and Regional Research* 32, no. 1 (2008): 1–21.

Treaster, J. "Police Quitting, Overwhelmed by Chaos." *The New York Times,* September 4, 2005.

———. J. DeSantis. "With Some Now at Breaking Point, City's Officers Tell of Pain and Pressure." *New York Times,* September 6, 2005.

UAE Yearbook. "Economic Development., 2007. http://www.uaeinteract.org/uaeint_misc/pdf_2007/English_2007/eyb5.pdf

UN Children's Fund (UNICEF). Annex II of S/1999/356, Section 18. 1999. http://www.un.org/Depts/oip/reports. February 17, 2004.

UN Habitat Program. *State of the World's Cities, United Nations Habitat Program*: Nairobi, 2007.

Urry J. *Global Complexity*. Cambridge, UK: Polity 2003.

———. *Mobilities*. Cambridge, UK: Polity, 2007.

U.S. Air Force. *Targeting, Air Force Strategy Document*. Washington, DC: U.S. Printing Office, 2006.

U.S.-Canada Power Outage Task Force. *Final Report on the August 14, 2003 Blackout in the United States and Canada: Causes and Recommendations* (April 2004). https://reports.energy.gov/BlackoutFinal-Web.pdf (accessed May 16, 2008).

———. *August 14th Blackout: Causes and Recommendations*, 2004.

U.S. Department of Defense. *Joint Vision 2020*. Washington, DC: U.S. State Department, 2000.

U.S. State Department. "Country Reports on Terrorism: Middle East and North Africa Overview." Office of the Coordinator for Counterterrorism, 2005. http://www.state.gov/s/ct/rls/crt/2005/64344.htm

Vale, L., and C., Thomas, eds. *The Resilient City: How Modern Cities Recover From Disaster*. Oxford, UK: Oxford University Press, 2005.

Van Houtum, H., and T. van Naerssen. "Bordering, Ordering and Othering." *Tijdschrift Voor Economische En Sociale Geografie* 93, no. 2 (2002): 125–36.

Van Wagner, E. "The Practice of Biosecurity in Canada: Public Health Legal Preparedness and Toronto's SARS Crisis." *Environment and Planning* A 40, no. 7 (2008): 1647–63.

Varnelis, K. "The Centripetal City: Telecommunications, the Internet, and the Shaping of the Modern Urban Environment." *Cabinet Magazine* 17 (Spring 2004). http://varnelis.net/articles/centripetal_city.

Ullman, E. "The Myth of Order." *Wired* (April 1999): 126–87.

Verbeek, P. *What Things Do: Philosophical Reflections on Technology, Agency and Design*. University Park: Pennsylvania State University Press, 2004.

Virilio, P. *The Art of the Motor*. Minneapolis: University of Minnesota Press, 1995.

———. *A Landscape of Events*. Cambridge, MA: MIT Press, 2000.

———. *Unknown Quantity*. London: Thames and Hudson, 2003.

Voigt, W. "From the Hippodrome to the Aerodrome, From the Air Station to the Terminal: European Airports 1909–1945." In *Building for Air Travel: Architecture and Design for Commercial Aviation*, edited by J. Zukowsky. Chicago: Prestel, 1996.

Wachtendorf, T. "Improvising 9/11: Organizational Improvisation Following the World Trade Center Disaster." Ph.D. diss., University of Delaware, 2004.

———. J. Kendra. "Improvising Disaster in the City of Jazz: Organizational Response to Hurricane Katrina." 2005. http://understandingkatrina.ssrc.org/Wachtendorf_Kendra. http://www.udel.edu/DRC/Wachtendorf_Improvising_9_11.pdf

Wald, M., R. Pérez-Pe–a, and N. Banerjee. "Experts Asking Why Problems Spread So Far." *New York Times*, August 16, 2003, A1, B4.

Walker, D. *For the Public's Health: Initial Report of the Ontario Expert Panel on SARS and Infectious Disease Control*. Toronto: Ministry of Health and Long-Term Care, 2004.

Wallace, W., M. Mendonca, E. Lee, J. Mitchell, and J. H. Chow. "Managing Disruptions to Critical Interdependent Infrastructures in the Context of the 2001 World Trade Center Attack." In *Impacts of and Human Response to the September 11, 2001 Disasters: What Research Tells Us,* edited by M. F. Myers, 165–98. Boulder, CO: Natural Hazards Research and Applications Information Center, University of Colorado, 2003.

Warden, J. "The Enemy as a System." *Airpower Journal* 9, no. 1 (1995): 41–55.

Warrick, J. "Crisis Communications Remain Flawed." *The Washington Post*, December 10, 2005.

Water Environment Federation. "Fat-Free Sewers: How to Prevent Fats, Oils, and Greases From Damaging Your Home and the Environment." (1999). http://www.wef.org/publicinfo/FactSheets/fatfree.jhtml

Weick, K., and K. Sutcliffe. *Managing the Unexpected: Assuring High Performance in an Age of Complexity*. San Francisco, CA: Wiley 2001.

Weick, K. "The Collapse of Sensemaking in Organizations: The Mann Gulch Disaster." *Administrative Science Quarterly* 38 (1993): 628–52.

Weizmann, E. *Hollow Land*. London: Verso, 2007.

Wenger, A., and R. Wollenmann. *Bioterrorism: Confronting a Complex Threat*. Boulder, CO: Lynne Rienner, 2007.

White House, The. *The National Strategy for Homeland Security*. 2002. http://www.whitehouse.gov/homeland/book/nat_strat_hls.pdf

Workforce Boards for Metropolitan Chicago (WBMC). *State of the Workforce, Report for the Chicago Metropolitan Region*. Chicago: WBMC, 2003.

World Health Organization. *Obesity: Preventing and Managing the Global Epidemic: Interim Report of a WHO Consultation on Obesity World Health Organization*. Geneva: WHO, 1997.

———. *SARS: Chronology of a Serial Killer* (Update 95, 2004). Geneva: WHO. http://www.who.int/csr

Wong, N., 1999. "The Role of the National Academies in Reducing the Vulnerabilities of Critical Infrastructures." Presentation to the Planning Meeting, National Academy of Sciences, Washington, DC, April 28–29, 1999.

Woodbridge, R. *The Next World War: Tribes, Cities, Nations, and Ecological Decline*. Toronto: University of Toronto Press, 2004.

Wright, A. "Structural Engineers Guide Infrastructure Bombing." *Engineering News Record*, April 3, 2003.

Yergin, D., and L. Makovich. "The System Did Not Fail. Yet the System Did." *New York Times,* August 17, 2003, 4–10.

Zaat, K. Norwegian Refugee Council (NRC)—Norway. "Isolation of Gaza Must End." Reuters, November, 29, 2007. http://www.alertnet.org/thenews/fromthefield/nrc/119632132787.htm

Zeitoun, M. "IDF Infrastructure Destruction by Bulldozer." *Electronic Intifada*, August 2, 2002.

Zimmerman, R. "Public Infrastructure Service Flexibility for Response and Recovery in the Attacks at the World Trade Center, September 11, 2001." In *Impacts of and Human Response to the September 11, 2001 Disasters: What Research Tells Us*, edited by M. F. Myers, 241–268. Natural Hazards Research and Applications Information Center, University of Colorado, Boulder, CO, 2003.

Zimmerman, R. "Social implications of infrastructure network interactions," *Journal of Urban Technology* 893 (2001): 97–119.

LIST OF CONTRIBUTORS

S. Harris Ali is a sociologist on the Faculty of Environmental Studies at York University, Toronto where he teaches courses on Environmental Disasters and Environment and Health. He has recently conducted research on disease outbreaks such as E. coli O157:H7 in Walkerton, Ontario and SARS in Toronto, Ontario. His most current work includes the volume, *Networked Disease: Emerging Infections and the Global City* (Wiley-Blackwell 2008) which was coedited with Roger Keil.

Deborah Cowen is an Assistant Professor in the Department of Geography and Program in Planning at the University of Toronto. Deborah is the author of *Military Workfare: The Soldier and Social Citizenship in Canada* (UTP 2008), coeditor with Emily Gilbert of *War, Citizenship, Territory* (Routledge 2008). Her work places the soldier at the center of the modern social world, examining how military models of work, discipline, domestic and urban space, and the social self shaped twentieth century welfarist and postwelfarist citizenship. She is currently investigating the transformation of logistics from military art to business science and so the rebordering of port cities, citizenship, and security.

Stephen Graham is Professor of Human Geography at Durham University, United Kingdom. He has a background in urbanism, planning, and the sociology of technology. His research addresses the complex intersections between urban places, mobilities, technology, war, surveillance, and geopolitics. He is Academic Director of the International Boundaries Research Unit (IBRU) and Associate Director of the Centre for the Study of Cities and Regions (CSCR), both at Durham. His books include *Telecommunications and the City*, *Splintering Urbanism* (both with Simon Marvin), the *Cybercities Reader,* and *Cities, War and Terrorism*. His latest book, *Cities Under Siege: The New Military Urbanism*, will be published by Verso in Fall 2009.

Roger Keil is the Director of the City Institute at York University and Professor at the Faculty of Environmental Studies at York University, Toronto. Among his publications are *The Global Cities Reader* (ed. with N. Brenner; Routledge 2006), *Networked Disease: Emerging Infections and the Global City* (ed. with S. H. Ali; Wiley-Blackwell 2008), and *Changing Toronto: Governing the Neoliberal City* (with J. A. Boudreau and D. Young; Broadview 2009). Keil's current research is on suburban infrastructure, cities and infectious disease, and regional governance. Keil is the coeditor of the *International Journal of Urban and Regional Research* and a cofounder of the International Network for Urban Research and Action.

Richard G. Little is Director of the Keston Institute for Public Finance and Infrastructure Policy at the University of Southern California where he teaches, consults, conducts research, and develops policy studies aimed at informing the discussion of infrastructure issues critical to California and the nation. Prior to joining USC, he was Director of the Board on Infrastructure and the Constructed Environment of the National Research Council (NRC) where he directed a program of studies in building and infrastructure research. He has conducted numerous studies dealing with life-cycle management and financing of infrastructure, project management, and hazard preparedness and mitigation and has lectured and published extensively on risk management and decision-making for critical infrastructure. Mr. Little has over thirty-five years experience in planning, management, and policy development relating to public facilities, including fifteen years with local government. He has been certified by examination by the American Institute of Certified Planners, is a member of the American Planning Association and the Society for Risk Analysis, and is Editor of the journal *Public Works Management and Policy.* He holds a BS in Geology and an MS in Urban-Environmental Studies, both from Rensselaer Polytechnic Institute. Mr. Little was elected to the National Academy of Construction in 2008.

Timothy W. Luke is University Distinguished Professor of Political Science at Virginia Polytechnic Institute and State University in Blacksburg, Virginia. He also serves as the Program Chair for Government and International Affairs in the School of Public and International Affairs, Director of the Center for Digital Discourse and Culture in the College of Liberal Arts and Human Sciences at Virginia Tech, and the University Senior Fellow for the Arts, Humanities, and Social Sciences. His research focuses on critical theory, environmental politics, and public culture.

Simon Marvin is currently the Lead Director of the Sustainable Urban and Regional Future (SURF) Centre, Salford University, and his research interests are focused on the codevelopment of cities and their sociotechnical networks. More recently his work has focused on understanding how new styles of urbanism are being developed in response to climate change and resource constraint which it is claimed can build ecological security for the world largest cities.

Colin McFarlane (PhD, Durham University) is Lecturer in Urban Geography at Durham University, United Kingdom. His research focuses on two interrelated themes developed through his work on Mumbai: geographies of infrastructure and the politics of knowledge, learning, and development. This has included work on the relationship between infrastructure, space, and inequality; the politics of 'slum' sanitation; and conceptualizing knowledge production and learning in translocal urban politics. He is currently writing a book based on doctoral and postdoctoral work entitled *Translocal Learning: Knowledge, Development and the North-South Divide* (Blackwell 2010). He has published on these themes in a range of academic journals.

Will Mead has a BSc (Hons) in Sociology and Social Policy (Southampton) and a PhD in Sociology (Lancaster). He is currently a Lecturer in Human Geography in Lancaster Environment Centre, Lancaster University, United Kingdom. The focus of his research is the social dimensions to sustainable water management for which he draws on literatures about everyday practice, sociotechnical systems, multilevel governance, socioecological resilience, urban political ecology, and complexity.

Benjamin Sims is a sociologist with the Statistical Sciences Group at Los Alamos National Laboratory. His research examines issues of risk and social order in relation to science and technology, with focus areas including infrastructure, laboratories, and nuclear weapons. He is primarily an ethnographer and qualitative analyst, and has worked in the areas of knowledge capture and expert judgment elicitation. His applied work has touched on a wide range of topics, including diaper manufacturing, terrorist decision making, and organizational issues in the U.S. nuclear weapons program.

Index

Page numbers in italic refer to Figures or Tables.